Learning Life Science in English

英語で読み解く 生命科学

寺本明子・James W. Hove・田所忠弘　編修

建帛社
KENPAKUSHA

刊行にあたって

　今，社会が大学生に求めているいわゆる「学士力」とはなにか？
　それぞれの狭い専門分野に留まることなく，地球全体にまでその思考や考え方を馳せることのできる人材が強く望まれているように感じます。国際社会への適応力・対応力の関係から，英語の力量アップも当然のことながら求められ，各大学ではさまざまな取り組みと工夫がなされています。
　一方，近年，「食料」，「環境」，「エネルギー」そして「健康」がキーワードとして注目されています。昨年3月に東北地方沿岸部を中心に襲った東日本大震災とそれに伴う原子力発電所の事故を契機に，これらはさらに胸中深く刻まれたことと思います。
　このような現況を背景として明るい未来を真摯に捉え，ライフサイエンス分野と英語分野の専門家の先生方に登場いただき，食料，環境，エネルギー，健康をテーマとして，それぞれの分野で学生にぜひ知っておいてもらいたい興味深い話題を英語版にて提供していただきました。
　科学知識の普及と英語による国際力の向上という2つの目標を合体させた本テキストで扱われるテーマは，自然科学分野だけではなく人文科学・社会科学分野の学生にも興味深く読んでいただけるものであると確信しています。『英語で読み解く生命科学』のタイトルどおり，普段は自然科学分野に接する機会の少ない学生にも，今や世界の共通言語としての英語をツールとし，英語の文章をとおして科学の一端に興味をもってもらえるよう，科学系の専門用語には解りやすい注解を付し，内容が理解しやすくなるよう編集しました。
　本書は，英語のテキストブックの体裁をとっておりますが，英語の文章をとおして各テーマの興味深さのみならず科学的な教養を高める一助とするという狙いももっています。一般の社会人の読者にもじゅうぶんに役立つ書籍として，やや専門的な用語解説も盛り込みました。英語のテキストということにこだわらず，教養や話題性を高める書籍として楽しんでいただければと思います。

　出版にあたり，ふさわしい話題の提供と，英語の文章をとおしたその理解に知恵を絞っていただいたライフサイエンス分野の先生方，英語の文章について適切なご指導・ご校閲と編集，さらには米国人からみた日本人の"食"の捉え方や考え方に興味深いテーマで切り込んでいただいたJames W. Hove先生，そして英語の文章をとおして科学の面白さに触れることができるよう編集にご尽力くださいました寺本明子先生に深く感謝申し上げるとともに，本書の刊行に尽力いただきました㈱建帛社の編集部の皆様，そして代表取締役社長の筑紫恒男氏に感謝申し上げます。

　2012年3月

編修者・著者を代表して
田所　忠弘

contents

- One American's Encounter with Eating in Japan and the Ongoing Discussion about Food
  …（日本人の食生活：ジェイムズ・W・ホーヴィー）……………………………… *1*

- Future Cheese …（未来のチーズ：佐藤 英一）…………………………… *9*

- The Hidden Power of Olives：Polyphenol
  …（オリーブの隠れた力：ポリフェノール：小林 謙一）…………………… *15*

- The Outside Within …（内なる外：小林 謙一）…………………… *21*

- Probiotics …（プロバイオティクス：佐藤 英一）…………………… *28*

- Controlling Hunger …（食行動はコントロールされている：山本 祐司）…… *35*

- Energy Metabolism and Metabolome Analysis
  …（エネルギー代謝とメタボローム解析：山本 祐司）………………………… *41*

- *Metabo*（Metabolic Syndrome）of "Food" and *Metabo* of "Information" …（「食」のメタボと「情報」のメタボ：小林 謙一）………… *46*

- Too Much Is as Bad as Too Little
  …（過ぎたるはなお及ばざるがごとし：小林 謙一）………………… *52*

- Cancer Is Complex …（癌は複雑である：山本 祐司）………………… *57*

- Researching the Global Environment at
  the Bottom of the Earth …（地球の底から環境を知る：沼波 秀樹）…… *62*

- Bioethanol …（バイオエタノール：佐藤 英一）………………… *69*

- Organisms that Eat the Earth
  …（地球を食べる生き物たち：沼波 秀樹）………………… *74*

- 科学解説 索引………………………………………………… *82*

One American's Encounter with Eating in Japan and the Ongoing Discussion about Food

1   "Dig in, mate," my vegetarian English friend said to me with an ear-to-ear grin on his face.

Three of my pals and I had been sitting in an all-night *izakaya* in Shinjuku, waiting for the train service to resume and take us back to our respective homes on the outskirts of Tokyo after a night on the town. Two in our group were vegetarians, the other was a Canadian who would have nothing to do with the "exotic" dishes of Japan, such as *sashimi*. So I had to wonder why the waiter was bringing a plate of fish to our table. I didn't order it. I looked down at the plate thrust in front of me to see a gutted fish, neatly filleted to be dipped into soy sauce laced with just the right amount of *wasabi*. This was not an unfamiliar sight to me after my year or so in Japan. But this time the fish's fins and tail were still twitching, and its gills were still gasping for air. Its eye was clear and staring straight up at me, or that's how it seemed. I shuddered a bit. The waiter remained at the table to see if I'd accept the dish. Since the thought of this fish's sacrifice being in vain horrified me even more than partaking of the flesh of something that was still moving, I uttered a timid "*Itadakimasu*." The waiter went back to work and I began eating. I thought I owed the fish, the chef, and the waiter at least that much. As I ate, I pondered just what I owed my vegetarian friend who had ordered this dish.

Even though the fish was quite tasty, this was my first and last experience with the practice of *ikizukuri* and my final initiation to Japanese cuisine. After a little over a year in Japan, I had pretty much sampled everything, even whale meat, and accustomed myself to the country's culture and customs of dining.

Having arrived in Japan in the late 1980s, the first major cultural hurdle I had to jump over was the simple act of going to the supermarket. I was living in Takamatsu on the island of Shikoku then, and my neighborhood food market was, compared to the massive supermarkets that resembled airplane hangars in the States, very tiny. It was just a shade bigger than the typical convenience store of today. And unlike the supermarkets of the United States where the smell of sanitizer was often stronger than the aromas of the food which I bought in colorful packages, aluminum trays, or plastic containers, the pungent smells of a wide variety of foods at the Japanese market were initially overwhelming. And where were the convenient packages of food? There were mostly fresh plants and parts of animals on the shelves and counter tops, and even in some cases things still moving around in tanks of water. Another nearby market had a tank that ran down its center where fish and other sea creatures would frolic until a customer would ask a clerk to fish one of them out. Needless to say, for the first few months my diet consisted mainly of familiar food items such as packaged *udon*, cup *ramen*, and the occasional recognizable cut of meat, which was still quite expensive back then.

After a few months I summoned up the courage and acquired the language ability to ask the fish merchant to pull a fish or some other sea creature from the tanks. I got a rice cooker and began to look forward to the seasonal fruits and vegetables to round off my meals. Certain realizations helped me get through the day more efficiently—a rice ball or two with a

One American's Encounter with Eating in Japan and the Ongoing Discussion about Food

suitable filling would provide more than enough energy to finish a long morning of work far better than the traditional bowl of cereal or the generously buttered slices of toast, egg and bacon of American breakfasts. My acclimation to Japanese eating practices also led to my own attempts at harmonizing Western and Eastern cuisine—filling a rice ball with eggs and bacon, for example. (Don't knock it 'till you try it.) And these experiments also gave me more to talk about with my Japanese colleagues when they invariably asked me what I had had for breakfast that morning.

And ask they did. It always surprised me just how much Japanese talk about food. Food and eating seemed to be such an integral part of the culture that I was hard-pressed to experience a moment of the day when someone was not discussing where, when, or what to eat. Turn on the TV and change the channels and you are almost sure to find a cooking program, a talk show that features cooking some traditional or exotic dish, or a drama in which the characters are enjoying a meal. What was all this fuss over food?

Until very recently in American history, food was traditionally viewed as nothing more than something to get you through the day, the fuel you needed to finish plowing the fields or help you drive the cattle across the plains. As the old cattleman John Wayne played in the 1972 movie "The Cowboys" summed this ethos up : "Slap some bacon on a biscuit and let's go! We're burnin' daylight!" Compared to their Japanese counterparts, Americans rarely eat in movies or TV shows, and, as any keen observer of American action movies and TV shows will note, the only one who ever does enjoy food is more often than not the villain. In Japan, however, everyone enjoys a delicious meal with a good conversation. The sociologist and critic Shimizu Ikutaro surmised that the constant eating in television and movies was due to the fact that the Japanese lacked the ability to appreciate fine conversation, yet I would assert that food is so central to the Japanese that it plays an indispensable role in their daily conversations. The anthropologist Takie Sugiyama Lebra takes one step further and contends that eating in Japan is an actual mode of communication.

Although Japan now grows and raises less than forty percent of its total food supply, you will find fruit and vegetables growing wherever you go in the country during almost every season and methods to prepare these unique to each region (*kyodo ryori*). It still never ceases to surprise me how common it is to come across rice fields or orchards in the midst of major cities. Take a ten-minute walk south of Ryoanji in Kyoto or a ride on the Chuo Line out to Mitaka in Tokyo and you'll find rice and vegetable fields with vending machines alongside them filled with the day's freshly picked produce. Where I live in the countryside in western Japan, you'll see flowers in bloom being frequented by birds and bees that pollinate the blossoms to bring *biwa* in the late spring, peaches in summer, *akebi* and persimmons in the fall and late fall. If you mention you buy vegetables at the supermarket, you may get a quizzical look : "You actually buy vegetables?"

If less so today, I still believe the Japanese have a keen awareness of what food is and where it comes from. The original meaning of the expression "*Itadakimasu*" said at the beginning of the meals is said to be a direct expression of gratitude to the fish, bird, or animal for the life given in order to nourish the person who eats it. One could make the case that saying grace in Western countries serves a similar function. However, in grace we Westerners thank God for providing us with the ability to put food on the table. Few, I would venture, ever make mention of the lives sacrificed for the meal. Perhaps this is why *ikizukuri* comes as such a shock to Western-

One American's Encounter with Eating in Japan and the Ongoing Discussion about Food

ers. The way most food is acquired in the West and the degree to which it has in most cases been processed has so totally dissociated people from the essential realities eating entails. What is cruel to Westerners may seem perfectly natural to people who are keenly aware that in order to sustain their own lives, other lives must be taken. I remember being taken slightly aback when I heard a group of Japanese people discussing how delicious the fish looked at an aquarium I was visiting. It seemed shocking at the time, but that's what a fillet of yellowtail or red snapper on your plate at a seafood restaurant essentially is.

What food is, what it should be, and how we can all improve our health and diet are just some of the more pressing matters we face among the multitude of problems that currently confront us. In America a major backlash against processed foods and the additives being introduced into them is brewing. Many books regarding this problem have topped best-seller lists, and more and more Americans are looking to other cultures for answers. And one of these cultures is that of Japan. In the recent American movie *Lost in Translation*, the main character realizes during a visit to Japan that "I want to take better care of myself. I would like to start eating healthier. I don't want all that pasta."

On the other hand, I have begun to notice the smell of sanitizer more and more in Japanese supermarkets. Very rarely do I find the variety of fresh vegetables I remember being confronted with in my first forays into Japanese food markets. And nothing on the shelves moves. More and more food advertised on TV and in magazines come in colorful wrappers, and you can find more and more ready-to-eat meals packed in plastic containers at Japanese supermarkets. As with their counterparts in the West, the number of Japanese households in which both parents are now working is on the rise leaving little time for meal preparation. The diversity of regional dishes found throughout Japan still exists, but it is questionable whether these meals are still being made in the homes, not in some far-away factory.

The popularity that Japanese cuisine boasts throughout the world and its image as a healthy alternative to certain aspects of the Western diet testifies to the fact that the Japanese have a lot to contribute to the discussion about what and how we will eat in the future. For the time being the ability to discuss food matters in English is the primary tool required to participate in the international discussion concerning how we will sustain not only our lives but those of the other living creatures with whom we share the planet. It is said that the more diverse the menu, the healthier the meal. Just as with a healthy meal, in the world's ongoing discussion on food and our future, the more people who add their voices to the conversation, the more fruitful and healthy it will prove to be.

英語 註

p.1

1	dig in：「食べ始める」
1	ear-to-ear：from ear to ear は，「耳から耳まで口にして」つまり，にやりと笑う様子。その形容詞形が ear-to-ear。
6	the other：ある集合の中で，〈何かを除いた残り全部〉と言う表現には定冠詞 the が必要。残りが一人（一つ）ならば，the other。残りが複数ならば the others。
6〜7	have nothing to do with 〜：「〜と関係が無い」 nothing の代わりに，something, much, little, a lot などが使える。
7	wonder：(wh- 節を後に付けて)「〜(なの)だろうか」
10	lace 〜 with…：「〜に…で風味を添える。〜に…を少量加える」
18, 19	owe：「恩義がある」 19 では「どのような仕返しをしようか」の意味。

p.2

2	sample：「試食する」
9	a shade：「ほんの少し」
17〜18	run down its center：「その中心にある」
19	needless to say：「言うまでもなく」
21	the occasional recognizable cut of meat：「たまに形を認めることができる程度の（小ささの）肉」
25	rice cooker：「炊飯器」
26	round off：「仕上げる，完成する」
27	rice ball：「おにぎり」

p.3

2〜3	generously butterd：「たっぷりバターを塗った」
6	Don't knock it 'till you try it.：「試してみるまでケチはつけないで」
12	where, when, or what to eat：「どこで，いつ，何を食べる」
15	What was all this fuss over food?：「いったいどうして食べ物のことでこんなに大騒ぎしているのだろうか？（と思った。）」
20	John Wayne：「ジョン・ウェイン」(1909-79 米国の俳優，西部劇映画のスター)
21	slap some bacon on a biscuit：「ぴしゃりとベーコンをビスケットの上に乗せる」
22	We're burnin' daylight!：「日が暮れるぞ！」
25	do：enjoy を強調
25	more often than not：「たいてい」
27	Shimizu Ikutaro：「清水幾太郎」(1907-88 日本の社会学者，評論家)

p.4

4	Takie Sugiyama Lebra：「タキエ　スギヤマ　リブラ」(1930- 日本生まれ，米国

	国籍の人類学者，日本研究家）
8	prepare：「調理する」
10	come across：「（偶然に）見つける」
11	ten-minute：「10分間の」
11	Ryoanji：「龍安寺」（京都にある寺）
14	produce：「農産物」
23	case：「主張」
24	grace：「（食前の）感謝の祈り」

p.5

3	entail：「伴う」
5	remember being taken slightly aback：「少し驚いたのを覚えている」
15〜16	look to other cultures for answers：「答えを求めて他の文化に目を向ける」
16	that：＝the culture
21	very rarely do I find：倒置
26	the number of 〜：〜の数　（a number of と混同しないように）

p.6

10	those：＝the lives

Future Cheese

1   Cheese is a food high in nutritional value and packed with calcium. There are various types of cheese throughout the world : Gouda in the Netherlands, Camembert in France, Emmentaler in Switzerland, and Mozzarella in Italy. In this article, however, I would like to talk not about cheese itself but
5   the way cheese is produced. More than that, this is an account of the future production of cheese utilizing gene recombination technology[*1]. In the methods of cheese production passed down through the ages since 4,000 BCE, cutting-edge biotechnologies now play a major role.

    To begin with, let's get an understanding of the steps involved in cheese
10 production : (1) Lactic acid bacteria[*2] are added to sterilized milk and fermented. (2) A milk-coagulating enzyme is then stirred into the milk. (3) Proteins in the milk aggregate and settle as curd. (4) The curd is placed in a box frame and excess water is removed. In this manner, the primitive cheese is formed. Delicious cheese with rich flavors can be produced
15 through further fermentation and ripening of the cheese. Incidentally, it is common to carry out this type of traditional cheese making in an open system. That is to say, cheese is made where airborne contaminant bacteria are present. But there's no need to panic : As lactic acid bacteria, the major fermenting enzyme in cheese, produce an antibacterial substance called
20 bacteriocin[*3], it's possible to produce hygienic cheese free from harmful

bacteria even within an open environment. A formidable foe to the lactic acid bacteria, however, is also present. It is a virus called bacteriophage (phage)[*4], which infects bacteria. Phages are the bane of cheese making that impede or terminate fermentation because they dissolve the lactic acid bacteria once they infect them. How to protect lactic acid bacteria from the threat of phage was a major dilemma. A number of techniques have been formulated so far, but here I would like to introduce you to a somewhat unconventional approach. This method makes use of a phage promoter[*5] (P_Φ) that becomes active only when a phage infects a bacterium. The deoxyribonuclease (DNase) gene should be coupled with the P_Φ promoter and integrated with lactic acid bacteria. So, when a lactic acid bacterium is infected with a phage, the P_Φ promoter turns on and starts to produce DNase. Since DNA is the blueprint on which essential information for a living thing to survive is encoded, lactic acid bacteria whose DNA is disrupted, of course, die. But, at the same time, since the phages accumulating inside these cells of the lactic acid bacteria become trapped and also die, only bacteria uninfected by phages are able to survive. This method is a touch extreme, but it is one system of defense against the phages you could refer to as a "strategy of mutual destruction."

Although I have already explained how phages are the bane of cheese manufacturers, from hereon I would like to introduce a strategy that turns the tables on these pests and puts them to use. But we'll first have to understand the process of cheese maturation a bit. The cheese produced through steps (1) through (4) mentioned above is called fresh cheese, which is still young cheese. When fermented further, proteins in this cheese are broken down into peptides and amino acids by peptidase[*6]. These peptides and amino acids give cheese its unique flavor and taste.

Peptidase can be found accumulated in the cells of lactic acid bacteria, and it is released through autolysis during the course of fermentation to help the cheese age over 3 to 6 months. Shortening this aging period brings economic advantage. That is where phages come in. Although phages are pests that dissolve lactic acid bacteria, a method that counters and takes advantage of this phenomenon to actively utilize the bacteriolysis caused by phages and to accelerate the aging of cheese has been developed. In this method, a bacteriolytic enzyme gene isolated from the phages is integrated with the lactic acid bacteria to artificially induce bacteriolysis. The timing of this bacteriolysis is crucial because it is problematic if it takes place indiscriminately. To prevent random autolysis, a system that strictly controls the activation of the bacteriolytic enzyme gene is required, and the method to induce bacteriolysis by using a special promoter was arrived at. In lactic acid bacteria, a promoter (P_{NaCl}) was discovered that activates only in the presence of sodium chloride at a certain concentration. If lactic acid bacteria can produce a bacteriolytic enzyme by using the P_{NaCl} promoter, the autolysis of lactic acid bacteria can be induced artificially by adding sodium chloride. As salt is added to age cheese during the actual process of cheese making, the idea to trigger autolysis through the addition of sodium chloride is an elegant solution. It is a novel idea to incorporate such a "timing device" into lactic acid bacteria to induce autolysis at an optimal time.

In addition to these, there is another case in which the generation of the peptide that causes the bitter taste of cheese is decreased. An excellent-tasting cheese can be produced by modifying through gene manipulation the area related to the substrate specificity[*7] of peptidase that generates the peptide producing a bitter taste and altering the pattern of decomposition of milk proteins to suppress the generation of this peptide.

1 In these ways, the use of gene recombination technologies reveals new possibilities for the process of traditional cheese making that has continued for thousands of years. Cheese utilizing the genetically-modified lactic acid bacteria introduced herein has not, however, yet hit the market. The reason
5 for this is that international consent regarding the use of genetically-modified food has not been reached. In Japan "genetically-modified microorganisms (food) safety criteria" have been established, and preparations for Japan taking a global initiative in this continue.

科学解説

*1　gene recombination technology ［遺伝子組換え技術］

　　　　細胞から取り出したDNAを試験管の中で人工的に操作し，その生物のもつ性質を改変したり，特定の遺伝子産物を作らせる技術のこと。遺伝子工学，組換えDNA技術とも呼ばれます。細菌だけでなく植物，動物にも応用することができ，バイオテクノロジーの最も基本的な技術です。

*2　lactic acid bacteria ［乳酸菌］

　　　　微生物分類学的にはグラム陽性，カタラーゼ陰性，運動性なし，胞子形成なし，消費したグルコースから50％以上の乳酸を生成する細菌の総称で学名ではありません。チーズ，ヨーグルト，漬物など多くの発酵食品製造の主要な発酵菌です。ヒトの腸内にも多く生息しており，整腸作用，免疫賦活作用など健康維持にも大きくかかわっています。

*3　bacteriocin ［バクテリオシン］

　　　　細菌が産生するタンパク質性あるいはペプチド性の抗菌性物質の一種。バクテリオシン産生菌と近縁な細菌にのみ抗菌効果を示すのが特徴で，この点が抗生物質とは異なります。チーズ発酵菌である *Lactococcus lactis* が産生するナイシンは食品添加物として日本を含む世界約50カ国で認可されています。

*4　bacteriophage ［バクテリオファージ］

　　　　細菌に感染するウイルスの総称で，正式には「バクテリオファージ」と呼ばれますが，「ファージ（phage）」という略称を使うこともあります。タンパク質でできた殻の中にDNAが閉じ込められただけの単純な構造をしています。ファージが感染し増殖するとファージが作る酵素によって細菌の細胞壁が溶かされてしまい，溶菌という現象を起こして死んでしまいます。

*5　promoter ［プロモーター］

　　　　遺伝子に書き込まれている情報からタンパク質を作るときのスイッチをON/OFFに調節する特別な配列。実際にはRNA合成酵素が結合する部位で，ここから遺伝情報がmRNAに写し取られ，それを基にタンパク質が作られます。

*6　peptidase ［ペプチダーゼ］

　　　　アミノ酸が多数（100残基未満程度）結合したペプチドに作用して，さらに分子量の小さいペプチドやアミノ酸に分解する酵素。ペプチドの端からアミノ酸・低分子量ペプチドを切りだすエキソペプチダーゼと，ペプチドの内部を切るエンドペプチダーゼの2種類があります。

*7　substrate specificity ［基質特異性］

　　　　ひとつの酵素は特定の物質のみを認識して反応を触媒します。これは酵素の活性中心という部位に特定の物質だけがはまり込むようになっているためで，他の物質とは反応できないようになっているのです。

英語 註

p.9

2〜3	Gouda, Camembert, Emmentaler, Mozzarella：「ゴーダチーズ，カマンベールチーズ，エメンタールチーズ，モッツァレラチーズ」
8	BCE：「紀元前」＝Before the Common [Christian] Era（非キリスト教徒によるB.C.に相当する記号）
8	cutting-edge：「最先端の」
10	bacterium：複数形は bacteria
11	milk-coagulating enzyme：「凝乳酵素」
15	ripening：「熟成」
17	that is to say：「つまり」
20	free from〜：「〜が無い」

p.10

6	a number of：「多くの」
7	so far：「これまでに」
7	introduce you to〜：introduce は第4文型を作れないので，to を忘れずに。
9〜10	deoxyribonuclease：DNA 分解酵素，デオキシリボヌクレアーゼ
17	a touch：「少し」
21〜22	turn the tables：「局面を逆転させる」

p.11

3	age：「熟成する」
3	over〜：「〜にわたって」
4	That is where phages come in.：「そこでファージの登場です」
6	phenomenon：複数形は phenomena
6	bacteriolysis：「溶菌」
8	bacteriolytic enzyme gene：「溶菌酵素遺伝子」
10	take place：「起こる，行われる」
15	sodium chloride：塩化ナトリウム
15	concentration：「濃度」
20	elegant：「すばらしい」
22	generation：「生成，発生」

p.12

4	hit the market：「市場に出る」

The Hidden Power of Olives : Polyphenol
—The Protective Mechanism of Plants
and Our Health—

1   Mention the word "olive," and most people will think of the olive oil used in Italian cuisine. However, the leaves of olives also contain useful substances. One of these substances is "oleuropein."[*1] It is one type of polyphenol[*2] and, as a matter of fact, functions as a biological protectant in olives. Furthermore, it has become evident that when we ingest it, oleuropein may act to protect our health and prevent metabolic syndrome. Let's consider what polyphenols are through a discussion of olives.

Think of the work of Vincent van Gogh, a representative artist of Post-Impressionism, and what will probably first come to mind is "Sunflowers." However, van Gogh also left many pictures with olive gardens as their subjects. Among them is the picture entitled "Olive Trees with Yellow Sky and Sun." It can be said that this picture is a typical van Gogh painting in which the shining sun is expressed through a yellow color and the olive trees are portrayed bathed in its rays. However, the olive tree is not a preferred subject of just van Gogh himself ; these pictures might also reflect how olives have been integral to European life since the eras of Greek civilization and the Roman Empire.

How can we appreciate this picture from a biological standpoint? Receiving sunlight, olive leaves photosynthesize and generate oxygen. Thanks to the oxygen they generate, animals, human beings included, are able to survive. However, reactive oxygen species (ROS)*³ is also generated during photosynthesis as a part of oxygen generation. ROS has a very strong oxidative power: the power to rust things. It is harmful not only to plants, but also to animals. In addition, strong sunlight may damage the genes of olives due to the ultraviolet rays contained in it. Furthermore, the warmth given off from the sunlight invigorates insects and bacteria, which come to olive trees to feed themselves. So, olive trees have to protect themselves from the harmful aspects of sunlight. If olives were animals, it would be enough for them to hide in the shade of trees or to go into a cellar to protect themselves from the sun's harsh rays. They could disperse insects and bacteria by simply brushing them off with their hands. Since olive trees cannot move and have no limbs, however, they must devise a technique to cope with these situations. Thus, they came to produce the substance oleuropein, a kind of polyphenol, as a defense mechanism.

As for the structure of the term "polyphenol," "poly," "pheno," and "ol" mean "many," a "benzene ring," and "alcohol (a hydroxyl group)" respectively. It is a complex chemical compound which consists of multiple benzene rings with hydroxyl groups. Due to this complex structure, polyphenols can take in ROS and protect a plant body from the bad influences of the contact with ROS. In addition, polyphenols have their own characteristic colors and serve like sunglasses to protect plants from ultraviolet rays. Since polyphenols also have peculiar tastes—most of which are bitter—they may act to keep insects and bacteria away from plants. It has been known that oleuropein also plays an active role, like other polyphenols, as a protective mechanism for plants.

What benefits, then, can we expect from these polyphenols when we consume them? Since we human beings have our own particular protective mechanisms by nature, we can neither produce polyphenols nor can we use them. Therefore, polyphenols exhibit various behaviors when we consume them, depending on their chemical property—some are absorbed directly into the body, some are absorbed after they are broken down by enterobacteria, and some are discharged. It is commonly said, that polyphenols mingle with nutrients and enter our body. However, even polyphenols that could fortunately manage to enter the body tend to be immediately expelled after the process of conjugation in the small intestine (referred to as enteroenteric circulation) or to be excreted together with bile after the process of conjugation in the liver (referred to as enterohepatic circulation). Conjugation is the reaction by which drugs and toxic substances are converted to non-toxic substances by being bound with glucuronic acid or amino acids, and are finally discharged. Thus, it seems that polyphenols are essentially treated as foreign substances in our body. Even so, some components of polyphenols that escape discharge probably act as antioxidants within our bodies as they do in plants. Oleuropein also acts as a potent antioxidant in our body and suppresses the oxidation of LDL cholesterol. It has been clarified that oleuropein suppresses the increase of blood glucose levels and enhances the capacity to generate heat in the brown adipose tissue ; it may play a role in preventing arteriosclerosis. Although olive leaves have been used as medicine or as an herbal tea since ancient times, the reason why consuming them is beneficial to human health cannot be explained. It may be possible that studies on oleuropein will elucidate this mechanism in the future. However, one thing should be emphasized : Polyphenols are functional substances that prove beneficial only toward the modern types of dietary habits that contribute to conditions of obesity.

The olive signifies "peace" in the language of flowers. Winners of marathons and other sporting events in ancient Greece were often presented a crown of olive branches and leaves. The words "olive branch" are in an idiom expressing "reconciliation" : "holding out the olive branch." Future study of olive leaves and oleuropein will prove them to be a food constituent that preserves the "peace" and promotes the "reconciliation" of substances within our bodies.

科学解説

生体内の抗酸化作用

　生体は，活性酸素の発生に対して防御機構を有しています。まず，スーパーオキシドは，スーパーオキシドジスムターゼ（superoxide dismutase：SOD）によって過酸化水素に変換されます。それによって，スーパーオキシドから毒性の高い活性酸素種であるヒドロキシラジカルが生じにくくなっています。また，SODによって生成された過酸化水素は，グルタチオンペルオキシダーゼ（glutathione peroxidase：GPX）の作用によって水にまで無毒化されます。それによって，過酸化水素から反応性の高いヒドロキシラジカルが発生するのを抑えています。また，GPXは，過酸化脂質を脂質アルコールへと変換するのにも関与しています。ほかにも，カタラーゼという酵素が，過酸化水素を水に変換しています。このような抗酸化機構が，活性酸素の発生とバランスをとっています。このバランスが崩れてしまい，活性酸素が多くなると，生体に毒性を示し，ガンや動脈硬化，糖尿病の発症と関連してくるのです。

＊1　oleuropein［オレウロペイン］

　　　オリーブに含まれるポリフェノールで，葉や実に含まれています。オレウロペインは，セコイリドイドに属しています。通常，オレウロペインは，糖が結合した配糖体として存在しており，消化の過程で糖が分解され，オレウロペインが分解されて生じたハイドロキシチロソールという形で吸収されます。体内で，抗酸化作用や抗炎症作用，そして血糖値上昇抑制作用や脂質代謝にも影響していることが明らかとなっています。褐色脂肪組織における熱産生を促進し，近年，肥満防止が期待されている成分もあります。

＊2　polyphenol［ポリフェノール］

　　　ポリフェノールとは，1つのベンゼン環に対して2個以上の水酸基（－OH基）をもつ化合物の総称です。通常は，単体（アグリコン）で存在しているのではなく，糖が結合した配糖体で存在しています。ポリフェノールは，植物由来であり，色素成分であったり，苦み成分であったり，紫外線から植物を守る成分であったりと，植物体内で重要な役割を果たしています。ポリフェノールは，ブルーベリーやブドウ，赤ワインに含まれるアントシアニンや，カカオ等に含まれるケルセチン，大豆などに含まれるイソフラボン，茶などに多いカテキン類など，種類も化学構造も非常に多様です。それらのほとんどが抗酸化作用を有しています。これらのポリフェノールが，ヒトの生体内で有効な機能を有することが明らかになり，注目されています。

＊3　reactive oxygen species（ROS）［活性酵素］

　　　本来，酸素がなければわれわれは生きてはいけません。栄養素を燃やしてエネルギーを得るために必要であるのは無論です。それだけではなく，酸素は，外部から侵入物を除去するときにも利用されています。しかし，酸素は多く存在すればよいかというとそうではなく，ものを錆びさせ，脆くさせることにもなりかねません。栄養素の燃焼も，ものが錆びることも，化学的には酸化反応という同じ反応なのです。したがって，多す

ぎる酸素は，かえって生体にとって悪影響を与えてしまいます。酵素分子の仲間には，ものを酸化させる能力がとりわけ強いものもあり，これらのことを活性酸素といいます。

英語 註

p.15
- 4　as a matter of fact：「実際，実を言うと」
- 4　function：「機能する」
- 8　Vincent van Gogh：「フィンセント・ファン・ゴッホ」(1853-90 オランダの後期印象派の画家)
- 8～9　Post Impressionism：「後期印象派」
- 11　among them is the picture：倒置
- 19　thanks to ～：「～のおかげで」

p.16
- 2　generation：⇒ p.11, 22 参照
- 6　bacteria：⇒ p.9, 10 参照
- 15　as for ～：「～に関して言えば」
- 16　hydroxyl group：「水酸基」

p.17
- 1　by nature：「本来」
- 1　nor can we use：倒置
- 4～5　enterobacteria：「腸内細菌」
- 8　conjugation：「抱合」
- 9　enteroenteric circulation：「腸腸循環」
- 10　enterohepatic circulation：「腸肝循環」
- 12　glucuronic acid：「グルクロン酸」
- 14　even so：「それはそうでも，それにしても」
- 15　antioxidant：「抗酸化物」
- 17　oxidation：「酸化」
- 19　adipose：「脂肪の」
- 19～20　tissue：「組織」
- 20　arteriosclerosis：「動脈硬化症」
- 25　prove ～：「～であることが判明する，分かる」

p.18
- 2～3　idiom：= "hold out the olive branch"「オリーブの枝を差し出す」で「和解の申し出をする」の意味

The Outside Within
—Tracing the Food We Eat to Make Clear What Digestion and Absorption Mean—

1   Suppose someone asks the question "Where do you think food enters the body when you eat something?" Most people would answer, "From the mouth, of course." But is this true? The answer would be "No." Well then, what could be the entrance for the food? (Where does the internal body actually begin?)

   The food we put into our mouth passes through the esophagus to the stomach, the small and the large intestines, and finally exits through the anus as feces. This series of organs is called the digestive tract. The digestive duct is basically a single tube. Can we then regard the inner wall of this tube, through which food passes, as an internal part of the body? The inner wall of the digestive duct opens to the external environment. Therefore, strictly speaking, it should be defined as an external part of the body. One way to consider this would be as "the outside within." If this is so, where does food enter the body in the true sense? This question can be answered by tracing the route of food after it is eaten.

   It goes without saying that we human beings are a kind of animal. Animals are living organisms that cannot directly utilize solar energy like plants, but they can survive by depriving other living things of their energy. Animals are thus destined to eat other living things. Therefore, the most important activity for animals is eating and, consequently, the most important organs are those used for eating. You may think that the brain is the most important to life, but in fact it originated from the organ for eating.

What is the central part of the body for eating? It is the digestive tract. It is a single tube from the mouth[*1] to the anus, including the esophagus[*2], the stomach[*3], the small intestine[*4], and the large intestine[*5]. The major characteristic of this series of the internal organs is that these organs are in contact with the external environment although they exist inside the body. They can therefore be called the area of "the outside within." In the digestive tract, living things taken as food are thoroughly decomposed through the ingenious process we call digestion, and the necessary constituents of the food are absorbed as nutrients[*6]. Unnecessary constituents are discharged.

Why must a living thing be thoroughly broken down to pieces? It is because food is not "food" from the viewpoint of things being eaten. When eating a fish, we are eating the living organism itself. When eating a leaf of a vegetable, we are eating the "place" for photosynthesis. When eating meat, we are eating an animal's muscles, tissue for movement. When eating an egg, we are eating a species' next generation, and its very cradle as well! Nothing we consider "food" intrinsically exists to be eaten. (The exception is "milk protein," which mothers produce to feed their children.) In other words, "food" carries the Purpose of the organism from which it originates : it carries the "information" intrinsic to that living thing.

Let's take protein[*7] as an example. Protein is a molecule made of numerous amino acids that are bound together by a peptide bond. In the natural world, there are 20 kinds of amino acids which can make up protein. Considering the possible bonds of these 20 kinds of amino acids to make a molecule of protein, it can be said that the combinations are almost infinite.

Furthermore, amino acids in protein are arranged not in a line but three-dimensionally, generating further structural variation. We can think of protein, then, as something in which "information" is written—as something with an intrinsic "design" itself.

What would happen if the protein we ingested remained in our body with its structure intact? If this happened, the "information" of the protein as something else, a "non-self," would enter the body and behave according to its "information." This would result in a disturbance of information in our body. To prevent such a disturbance, a process of digestion becomes necessary, in which the "information" of food, the "non-self," is broken down into the universal "information" all living things can assimilate.

Since the digestive tract is a place where the "non-self" and "self" compete with each other, a tight guard is required to prevent the disturbance of the "information" in the body. This guard is what we refer to as immunity. Immunity is the process in our body that distinguishes the "information" of the "self" from that of the "non-self," and acts to thoroughly exclude things related to the latter. Although you may think that this process only occurs within the bone marrow or blood, the digestive tract is also where many immunity-related cells are placed. The Peyer's patch is an example of cells bearing such a role. That is, it cannot be allowed to take in the "information" of the organism "eaten".

Since the digestive duct is the internal system for eating, it needs to break down living things efficiently to select only constituents required for the body and to assimilate them into the body. It is the nerve cells that carry out a role of a "control board" to regulate the transportation and digestion speed of food in the digestive tract. They always monitor the quality and quantity of food passing through the digestive tract.

1   The mucosal intestinal epithelial cells, where "absorption" takes place, work very hard. They have to absorb nutrients as well as digest food. They are located at a point of contact with the external environment and can always come into contact with foreign substances. Therefore, they will be too
5 exhausted to perform unless there is a rapid turnover. For this reason, the lifetime of the mucosal epithelial cells can be as extremely short as a few days, and the new small intestine cells are produced steadily while old ones are shed off.

   When we think scientifically about the meaning of eating through the
10 study of the digestive tract, we can be led to ponder the philosophical proposition, "the outside within."

科学解説

内なる外

消化管腔は，厳密には外部に接しているがゆえに「外部」です。外部に接していない部分が「内部」です。したがって，血管内は，「内側」とみることができます。血液中のホルモンのことを内分泌ということがありますが，それは，ホルモンが体内で合成され，血液中に分泌され，体内で作用を発揮するから，「内分泌」といわれているわけです。逆に，汗や涙は，体外に分泌されるので，「外分泌」といいます。消化器系でも，消化管腔に分泌される唾液や胃液，膵液などは，厳密には「外部」に分泌されているという理由で「外分泌」に分類されています。

＊1　mouth［口腔］

　　　食品がまず通過する消化管です。口腔で，歯によって咀嚼され，食品が物理的にちぎられます。その間に食品は，唾液が混合され，唾液中のアミラーゼやリパーゼによって一部の分解が起こり，粘液によって食塊が消化管を通過しやすくなります。

＊2　esophagus［食道］

　　　食塊は，舌の運動によって飲み込まれ食道に送り込まれます。これを嚥下といいます。食塊は，食道の蠕動運動によって胃へ送られていきます。

＊3　stomach［胃］

　　　消化管の中で最も拡張している部分であり，成人で容積が1.2〜1.4リットルになります。胃の入り口は噴門といい，出口は幽門といいます。胃の内壁はすり鉢状になっており，蠕動運動によって，胃液と混合された食塊は，半流動の粥状のキームスとなります。胃液には，胃酸（塩酸）とペプシンと呼ばれるタンパク質分解酵素が含まれています。食品中のタンパク質は，胃酸による変性作用とペプシンにより分解されます。胃は，小腸における本格的な消化作用の準備を行う場といえます。

＊4　small intestine［小腸］

　　　糖質，脂質，タンパク質の消化・吸収の大部分が行われる部分です。小腸は，大きく3つの部分に分かれ，胃の幽門に続く部分を十二指腸，それに続く空腸，残りの部分を回腸といいます。それぞれの境目は明確ではありません。小腸の内壁は，ひだ状になっており，その表面は絨毛と呼ばれています。絨毛の表面は，小腸粘膜上皮細胞が整然と配置されています。小腸粘膜上皮細胞の表面は微絨毛と呼ばれるブラシ状の構造となっており，表面積が最大になるようになっています。

＊5　large intestine［大腸］

　　　小腸に続く部分をいい，盲腸，結腸，直腸に分かれます。大腸内で，内容物中の水分や無機物が吸収されていきます。また，大腸内には腸内細菌が棲息しており，発酵による未消化物の分解が起こります。そこから得られたビタミンなどが体内に吸収されます。このような消化のことを生物学的消化といいます。

*6　nutrient［栄養素］

　　生物が生命現象を行うためには，絶えず外界から物質を取り込み，生体内で代謝し，不要なものを体外に排出する必要があります。これらの一連の状態を栄養（nutrition）といいます。また，生物が生命現象を行ううえで必要不可欠で，体外から取り入れなければならない物質のことを栄養素（nutrient）といいます。栄養素となる物質は，生物種によって異なり，ヒトの場合，栄養素は食品中に含まれ，糖質（炭水化物），脂質，タンパク質，無機質（ミネラル），そしてビタミンの5つ（5大栄養素）に分けられます。

*7　protein［タンパク質］

　　タンパク質は生物を構成する主成分であり，生命現象を行うさまざまな反応に重要な役割を果たしている，窒素を含む高分子化合物です。タンパク質は，アミノ酸（タンパク質を構成するアミノ酸は約20種類）がペプチド結合で数珠つなぎになったものです。天然に存在するタンパク質は，それを構成するアミノ酸の種類，数，並び方がすべて決まっています。アミノ酸の並び方のことをアミノ酸配列（1次構造）といいます。アミノ酸配列を決めているのが遺伝子と呼ばれるもので，タンパク質の設計図の役割を果たしています。また，タンパク質のアミノ酸配列のなかでも，アミノ酸の側鎖と呼ばれる部分同士が結びつき，らせん状の構造やシート状の構造になることを2次構造といいます。さらにそれらを総合すると，ひとつのタンパク質は，2次構造や不規則な構造を含む，独特な立体構造をとることになります。このことを3次構造といいます。3次構造をとるタンパク質が複数集まり構造体を形成することを4次構造といいます。さまざまな複雑な生命現象を可能にするのは，タンパク質の立体的な「形（かたち）」なのです。

英語 註

p.21

8	series：	a series of 〜は，「一連の〜」という意味だが，単数扱い。
8	digestive tract：	「消化管」
12	strictly speaking：	「厳密に言えば」
16	it goes without saying that 〜：	「〜は言うまでもない」

p.22

14	from the viewpoint of things being eaten：	「食べられるものの観点から見ると」
17	tissue：	⇒ p.17, 19〜20 参照
18	species：	「種」単複同形
18	generation：	「世代」
18	very：	「まさにその」
23	molecule：	「分子」
24	peptide bond：	「ペプチド結合」

p.23

16   that：＝the information

17   the latter：「後者」＝the information of the "non-self"

18   bone marrow：「骨髄」

19   Peyer's patch：「パイエル板」

p.24

1   mucosal intestinal epithelial cell：「小腸の粘膜上皮細胞」

1   take place：⇒ p.11, 10 参照

7   ones：＝small intestine cells

Probiotics

1   The richly diverse intestinal microbiota*[1] consists of as many as 100 trillion bacteria of more than 1,000 species that inhabit the human digestive tract. These bacteria produce various substances through their metabolic processes to maintain their lives, and our health is acutely affected by the
5   behavior of this intestinal microbiota since we human beings as the host of these bacteria receive their metabolites as irritants. It is also known that these enterobacteria themselves deeply affect the maintenance of our health. Here I would like to give an account of the helpful bacteria we refer to as probiotics.

10 The influence that this massive population of organisms has on their host has garnered attention to establish the concept of "probiotics": the expectation of health-promoting effects for the host through the active ingestion of useful bacteria to invigorate the intestinal microbiota. "Probiotics" is a term that contrasts with "antibiotics" and originates from an ecological
15 term meaning symbiosis between different species of organisms, and it is expected to be highly safe and exhibit evidence of healthy benefits for the body. Examples of such probiotics are so-called "beneficial bacteria" as lactic acid bacteria and bifidobacteria, which have been incorporated into the daily diets of people all over the world in such products as yogurt.
20 In 1904, the Russian microbiologist and zoologist Ilya Ilyich Mechnikov

advocated his theory for perennial youth and the prolongation of life, claiming that intestinal decay caused premature aging and death and that aging could be prevented by actively consuming sour milk such as yogurt because the lactic acid bacteria contained in it eliminates putrefying bacteria from the intestine. However, this theory fell by the wayside since the opinion that lactic acid bacteria contained in yogurt could not adhere to the intestinal tissue and subsequently ended up dying in the body became the established norm. Nevertheless, the Mechnikov article is regarded as the earliest report on the health effects derived from fermented milk. And now, a century after this report, a lot of evidence for the health benefits from ingesting lactic acid bacteria has been reported. And these are just some of those examples : reduction in the risks of allergies ; alleviation of inflammatory bowel disease[*2] ; adjuvant effects[*3] ; and phylaxis against pathogenic bacteria.

Among these benefits I'd first like to briefly describe their effects on allergies. Human beings produce antibodies[*4] internally to repel allergens[*5] when those from pollen or ticks invade the human body. Healthy people usually react to these allergens involuntarily as a conditioned response. But from a certain point in time this conditioned response can suddenly become an allergic reaction toward allergens and produce huge amounts of antibodies. An inflammatory reaction consequently takes place at the sites where these antigens make contact. This abnormal reaction is apt to occur when the balance of the helper T lymphocytes[*6] is disturbed. There are two types of helper T lymphocytes : Th1 and Th2. Both types of lymphocytes reciprocally maintain balance to control the immune response, but for some reason Th2 lymphocytes become excessive in people who suffer from allergies. As a result, the level of antibodies in the blood rises in patients with allergies as their production against allergens, such as pollen and ticks, is triggered by the Th2 lymphocytes. It is thought that the aggravation of allergic diseases, such as atopic dermatitis and hay fever, can be impeded by controlling the increase of Th2 lymphocytes, i.e., improving the balance of Th1 and Th2 lymphocytes. It has recently been reported that lactic acid bacteria regulate the distribution of the helper T lymphocytes, shifting the balance of lymphocytes from Th2 to Th1. It has come to light, however, that this effect varies with respective strains even among the same species. Therefore, it is now considered necessary to ingest a bacterial strain to suit each symptoms.

Moreover, a process for "beneficial bacteria" to normalize the intestinal micorobiota is required to bring about beneficial effects : stages in which the beneficial bacteria adhere to the intestinal walls to establish a foothold and act on the surrounding micro biota and host tissue. The adhesion mechanism of lactic acid bacteria, however, has not been fully clarified in spite of accounts of their beneficial effects. This owes mainly to the fact that the safety of lactic acid bacteria, unlike pathogenic bacteria, has been empirically verified. When pathogenic bacteria adhere to intestinal tissue, they produce such afflictions as diarrhea, abdominal pain and vomiting, and may prove fatal should worse come to worst. As pathogenic bacteria produce these serious afflictions, the processes in which they exert such harmful effects continue to be thoroughly investigated to this day. Since the safety of the bacteria commonly referred to as "beneficial bacteria", such as lactic acid bacteria, has been confirmed, the mechanism for their adhesion to the intestinal lining has contrarily been neglected. However, since the beneficial effects discussed above have been proven scientifically in recent years, the explication of the adhesion mechanism of lactic acid bacteria is being pursued in order to bring out its beneficial effects more efficiently. Research so far has revealed that two types of lactic acid bacteria exist : those that adhere easily and those that do not.

Furthermore, an oral vaccine[*7] that corresponds to these phenomena is now being developed. Briefly, this oral vaccine is a revolutionary technology that establishes immunity not through parenteral administration such as injection but the oral ingestion of an immunogen. This method carries a detoxified protein, the antigen, to the digestive tract through lactic acid bacteria to effect immunity. Although this method is currently in the developmental stage, the time may come when injections may become obsolete. This type of research on probiotics continues to flourish, and the benefits of time-honored fermented foods, such as yogurt, continue to be clarified through modern scientific methods. Although much in this field yet remains unknown, this area of research can contribute to human health, and is sure to make further strides toward future food products offering greater benefits.

> **科学解説**

*1　intestinal microbiota ［腸内細菌叢］

　　　　ヒトや動物の腸の内部に生息している細菌のことを指します。宿主であるヒトや動物が摂取した栄養分の一部を利用し，他の種類の腸内細菌との間で数のバランスを保ちながら独特の生態系を形成しています。種類と数は，動物種や個体差，消化管の部位，年齢，食事の内容や体調によって違いがみられますが，その大部分は偏性嫌気性菌です。大腸菌など培養可能な種類は全体の一部であり，大部分の菌はいまだ明らかにされていません。

*2　inflammatory bowel disease ［炎症性腸疾患］

　　　　長期に下痢，血便が続く原因不明の難病です。通常の食中毒などと異なり，数日でよくなることはなく長期にわたります。適切な治療を行えば通常の生活をおくれますが，完全に治ることはありません。命を落とすことはありませんが，病気のために生活が大きく犠牲になるのがこの病気の特徴です。

*3　adjuvant effect ［免疫賦活効果］

　　　　「アジュバント」は元来「助ける・振動する」という意味ですが，免疫学で用いられる場合，抗原と混合または組み合わせることで抗体産生の増大，免疫応答の増強を起こす物質の総称です。また，NK（ナチュラルキラー）細胞など，異物に対して中心となって闘う免疫細胞を活性化させるため，癌を予防・抑制したり，風邪などの感染症を未然に防ぎ，回復を早めたりする効果があります。

*4　antibody ［抗体］

　　　　抗体は体内に侵入してきた細菌・ウイルスなどの微生物や，微生物に感染した細胞を抗原として認識して結合するタンパク質です。抗原へ結合すると，複合体を食細胞が認識して体内から排除するように働き，免疫細胞が結合して免疫反応をひき起こしたりします。

*5　allergen ［アレルゲン］

　　　　アレルゲンとはアレルギー疾患をもっている人の抗体と特異的に反応する抗原のことを指します。一般には，そのアレルギー症状をひき起こす原因となるものをいいますが，健常な人においても，その抗体と反応する抗原もアレルゲンと呼びます。さらに広義には，それに対するアレルギー患者が多いなど，アレルギーの原因によくなりうる物質のことをいいます。

*6　helper T lymphocyte ［ヘルパー T 細胞］

　　　　ヘルパー T 細胞は，体内で見つける細胞が自分自身か，非自己かを見分ける免疫細胞で，抗原を敵として認識し攻撃します。さらに，マクロファージから抗原の情報を受け取り，β 細胞に抗体を作るよう指令を出し，β 細胞が抗体を作るのを助け，マクロファージが活性化するのを助けます。ちなみにエイズ・ウイルスは，ヘルパー T 細胞についてのみ特異的に感染して T 細胞を破壊し続けるウイルスです。

*7　oral vaccine［経口ワクチン］
　　　　現在のワクチン投与法は注射による皮下投与が主流ですが，この方法は血中抗体が免疫の主体となり，粘膜局所免疫はほとんど誘導されません。多くの感染症は病原体が粘膜表面へ接触し，粘膜から侵入するか，またはその部分で定着増菌してはじまります。したがって，口から投与したワクチンを粘膜まで運んでやれば粘膜上の粘膜局所免疫系を刺激して免疫を誘導させることができます。この免疫は局所免疫ばかりでなく全身系の免疫をも成立させることができるのです。

英語 註

p.28

2～3	digestive tract：p.21, 8 参照	
6	metabolite：「代謝産物」	
6	irritant：「刺激物」	
7	enterobacteria：p.17, 4～5 参照	
14, 15	term：「用語」	
14	antibiotics：「抗生物質」	
15	symbiosis：「共生」	
17	so-called：「いわゆる」	
18	bifidobacteria：「ビフィズス菌」（ビフィドバクテリウム属に属する細菌の総称）	
20	Ilya Ilyich Mechnikov：「イリヤ・イリイチ・メチニコフ」（1845-1916 ロシアの微生物学者，動物学者）	

p.29

5	fall by the wayside：「挫折する，忘れ去られる」	
7	tissue：p.17, 19～20 参照	
7	end up - ing：「結局 - することになる」	
13	phylaxis：「（感染）防御」	
13	pathogenic：「病原（性）の」	
17	those：＝ the allergens	
18	conditional response：「条件反射」	
21	take place：p.11, 10 参照	
22	antigen：「抗原」	

p.30

3	atopic dermatitis：「アトピー性皮膚炎」	
3	hay fever：「花粉症」	
4	i.e.：＝ id est「すなわち」	
8	strain：「菌株」	

12	bring about：「もたらす」	
16	owe to ~：「~に帰せられるべき」	
19	diarrhea：「下痢」	
19	abdominal：「腹部の」	
20	prove：p.17, 25 参照	
20	should worse come to worst： = if worse should come to worst「万一最悪の事態になったら」	

p.31

2	so far：⇒ p.10, 7 参照	
3	those： = the lactic acid bacteria	
6	parenteral：「非経口的な」	
6	administration：「投与，施薬」	
7	immunogen：「免疫原」	

Controlling Hunger

The act of eating is strictly controlled: just being hungry does not arouse your "appetite"; neither does simply being full cause you to lose it. Appetite is regulated by the brain—something akin to a control tower exists in the "appetite center"[*1] of the hypothalamus. The "autonomic nerve center" resides here, and the autonomic nervous system maintains control of appetite. These "autonomic nerves" are found in the "sympathetic nervous system" and the "parasympathetic nervous system." The sympathetic nervous system functions in a "state of tension," whereas the parasympathetic nervous system operates in a "relaxed state." This is the reason, for example, you find that when you are tense before an examination, you don't have much of an appetite, but once it's over and you begin to "relax," your appetite returns. Here, I would like to succinctly illustrate how metabolism and its related hormones[*2] act on the brain to modulate appetite.

Metabolism (biological activity) is the balance between catabolism and anabolism. That is to say, vital phenomena mean the balance between synthesis and decomposition. It has recently come to light that metabolism also acts as an important cuing factor in the control of appetite. For example, when we are hungry, our blood glucose level[*3] starts to drop and the level of free fatty acids[*4] rises. It is known that changes in the levels of these humoral factors are sensed by the brain, the sympathetic nervous system activates, secretory organs excrete hormones into the circulatory system, and all of these actions are involved in homeostatic maintenance. During hypoglycemic states, to take an instance, the glucagon are secreted from the pancreas, and adrenalin (epinephrine) as well as glucocorticoids are secreted from the adrenal gland, in order to raise the level of blood glucose. These hormones affect the cells of organs to fundamentally act in the increase of level of blood glucose. Next, glucagon binds with receptors on cell membranes to hasten gluconeogenesis[*5]. It also stimulates glycogenolysis and spurs glucose production. The concentration of ATP[*6] also drops because the generation of fresh ATP within cells no longer takes place and the ratio of AMP to ATP elevates. It is known that such instances activate AMPK[*7], which is an intracellular signaling factor that detects the change in the ratio of AMP to ATP. Indeed, it has become evident from the results of experiments using rats that the artificial activation of AMPK through drugs induces an increase in appetite. In addition, it is known that AMPK functions as a glucose sensor in the hypothalamus of the brain. Ghrelin is a peptide hormone made up of 28 amino acids that also stimulates the appetite, and it is secreted by the stomach to relay signals to the brain through the vargus nerve that has a profound relation on the parasympathetic nervous system. The level of ghrelin in the blood increases prior to meals and promptly drops after meals. The discovery of ghrelin somewhat accounts for the mechanism in which the stomach is able to control the action of eating.

When we consume a meal, on the other hand, blood glucose levels increase and free fatty acid levels decrease. The parasympathetic nervous system is consequently stimulated, and the pancreas secretes insulin. Insulin acts on hepatic and muscular tissue to enhance glucose intake and to stimulate glycogen synthesis, causing the level of glucose within the blood to return to normal. In this manner, insulin has come to be recognized as the only hormone that decreases high levels of glucose within the blood. Insulin secretion is controlled not only by the parasympathetic nervous system but also by a variety of other signals—one of which is incretin, a gastrointestinal hormone released from the digestive tract after ingestion of a meal to stimulate insulin secretion from the pancreas. Interestingly enough, it has also come to light that the ghrelin secreted from the stomach suppresses insulin secretion. This is another case in which a hormone controls appetite and metabolism. What's more, leptin is secreted from adipose cells. The leptin gene is a hormone encoded with an obese gene (*ob* gene)[*8], and it was found by Friedman, et. al. of Rockefeller University in the United States. It also became evident that there is a receptor of leptin in the hypothalamus to suppress appetite. Here ghrelin influences the hypothalamus as an antagonistic hormone against leptin, apparently to control appetite and metabolism as well. Recent studies indicate the prospect of insulin directly acting on nerves in the hypothalamus to control appetite, suggesting the diverse effects of insulin.

Further, it has become clear that cholecystokinin (CCK), a hormone secreted from the small intestine into the blood, not only stimulates the secretion of digestive enzymes but also suppresses appetite. On top of all of this, it is known that nervous histamine[*9], which is synthesized from histidine, a kind of amino acid, in the brain, suppresses hunger via the histamine receptor. It was reported that the administration of histidine suppressed appetite in rats, but it does not seem feasible to regard it as an ingredient to suppress appetite on account of its bitter taste. Not only that, α-lipoic acid[*10] was also reported to suppress appetite, indicating the growing possibility that appetite may one day be controlled through food ingredients and sparking hope for further progress in this type of research in the future.

食行動はコントロールされている

> **科学解説**

*1　appetite center［食欲中枢］

　　　食欲は，脳内の視床下部に存在する食欲中枢（摂食中枢と満腹中枢）によってコントロールされています。

*2　hormone［ホルモン］

　　　ホルモンは臓器から分泌される生理活性物質です。ヒトの体には恒常性を保つような仕組みがあり，そのバランスが崩れると病気になります。ホルモンはそのバランスを保つようにシグナル因子として機能して，代謝などをコントロールしています。ホルモンには油に溶ける脂溶性ホルモン（ステロイドホルモンなど）と油に溶けない水溶性ホルモン（インスリンなどのアミノ酸などにより構成されるもの）があります。

*3　blood glucose level［血糖値］

　　　ヒトの血液中の糖，すなわちグルコース（$C_6H_{12}O_6$）の濃度は空腹時で約 100 mg/dL となっています。これは脳がグルコースを唯一のエネルギー源（車のガソリンと考えてみてください）としているためであり，常に脳に安定的にエネルギー源を供給するために，この濃度より下がらないようになっています。

*4　free fatty acid［遊離脂肪酸］

　　　一般にヒトが食べる脂質には中性脂肪，コレステロール，リン脂質などがあります。遊離脂肪酸は中性脂肪やリン脂質を構成している化合物であり，糖（グルコース）とともにエネルギー源としての役割があります。血糖値が低下すると脂肪組織の中性脂肪が分解されて遊離脂肪酸が上昇して，生命活動に必要なエネルギーを供給します。

*5　gluconeogenesis［糖新生機構］

　　　⇒ p.44　科学解説 *7 参照

*6　ATP［アデノシン三リン酸］

　　　adenosine triphosphate の略称で，体の中の高エネルギー化合物のひとつです。体を動かす電池のようなものと考えてください。1分子の ATP からリン酸がひとつ取れると約 7 キロカロリーの熱エネルギーを放出し，アデノシン二リン酸（ADP）となります。ADP にリン酸を再び結合すると ATP が産生されます。リサイクル型の電池です。ただ，リサイクルするのにもエネルギーが必要ですので，グルコースや遊離脂肪酸を分解してそのエネルギーを供給してもらうのです。

*7　AMPK［AMP 依存性キナーゼ］

　　　AMPK は細胞中の ATP/AMP 比を感知して，それを信号として細胞内に伝えるタンパク質です。つまり，エネルギーセンサーと考えてください。

*8　obese gene（*ob* gene）［肥満遺伝子］

　　　米国の Friedman らが，長い時間をかけて太ったマウスをかけ合わせることで遺伝的な肥満マウスを作成し，その原因遺伝子を同定して発見された遺伝子です。脂肪細胞から分泌される食欲抑制ホルモンであり，遺伝的に欠損すると過食により太ることが解っ

＊9　nervous histamine［神経ヒスタミン］

　　　　ヒスタミンというとアレルギーを連想しますが，脳内でもヒスチジンというアミノ酸からヒスタミンが生合成されています。これを神経ヒスタミンと呼び，食欲中枢に働きかけて食欲を抑制することが解ってきました。

＊10　α-lipoic acid［α-リポ酸］

　　　　酵素の働きを助ける有機化合物で，エネルギー産生にかかわる代謝経路を活性化することが知られています。したがって，AMPKを介して食欲抑制効果を発揮するのではないかと考えられています。

英語 註

p.35

行	英語	意味
2	full	：「満腹の」
4	hypothalamus	：「視床下部」
4〜5	autonomic nerve center	：「自律神経中枢」
6〜7	sympathetic nervous system	：「交感神経系」
7	parasympathetic nervous system	：「副交感神経系」
14	catabolism	：「異化作用」
15	anabolism	：「同化作用」
15	that is to say	：⇒ p.9, 17 参照
15	phenomenon	：⇒ p.11, 6 参照
15〜16	synthesis	：「合成」
16	decomposition	：「分解」
20	humoral factor	：「液性因子」
21	secretory organ	：「分泌器官」
21	circulatory system	：「循環系統」
22	homeostatic	：「恒常性の」
23	hypoglycemic	：「低血糖の」
23	to take an instance	：「一例を挙げると」

p.36

行	英語	意味
1	pancreas	：「膵臓」
1	glucocorticoid	：「グルココルチコイド」
2	adrenal gland	：「副腎」
4〜5	cell membrane	：「細胞膜」
5	glycogenolysis	：「グリコーゲン分解」
7	take place	：⇒ p.11, 10 参照

16	vagus nerve：「迷走神経」	
18〜19	account for 〜：「〜を説明する」	
21	consume：「摂取する」	
24	hepatic：「肝臓の」	

p.37

3	digestive tract：⇒ p.21, 8 参照	
7	what's more：「さらに」	
7	adipose：⇒ p.17, 19 参照	
9	et.al.：「およびその他の者」	
12	antagonistic hormone：「拮抗ホルモン」	
16	cholecystokinin（CCK）：「コレシストキニン」	
18	on top of 〜：「〜に加えて」	
21	administration：⇒ p.31, 6 参照	
23	on account of 〜：「〜の理由で」	

Energy Metabolism and Metabolome Analysis

You could cite self replication (heredity) and the capability to synthesize and break down substances in cells (metabolism) as the most important functions to life. In our body, homeostasis is maintained through our metabolism, namely, the regulation of catabolism and anabolism. In higher organisms such as human beings, hormones control the metabolism as well as the proliferation and differentiation of cells—and what actually is responsible for regulating these processes is the activity of enzymes. To be more precise, the genetic information managing the information of proteins that form enzymes is thought to control the metabolism. Higher organisms additionally maintain normal functions not only through the supply of nutrients but also through the exchange of information between organs and cells through blood. Consequently, blood circulation is considered to be the information medium that networks organ to organ, cell to cell.

When glucose is absorbed into the body from a meal, the concentration of glucose[*1] in the blood elevates. This level of blood glucose is regulated to remain constant. Excess amounts of glucose in the blood are drawn into the liver and muscle to become glycogen[*2], the polymerized form of glucose, and stored until it is required. In addition, lipid (especially neutral lipid[*3]) is drawn into the liver and adipose tissue to be stored. Assimilated nutrients are not only stored in their original form but are also broken down through catabolism, and the energy produced during catabolism is converted to high-energy compounds that can be used when the need arises, and then stored. Since energy resources in the body, such as glucose and neutral lipids, are in a reduced[*4] state, they release energy into the body when they are oxidized[*4] through the process of catabolism. As neutral lipid is in a more highly reduced state than glucose, it releases an energy amount of 37 kJ/g calories, whereas glucose releases only 16 kJ/g calories. Since neutral lipid can supply energy in a more concentrated form than glucose, energy is mainly stored as lipid in the adipose tissue. The reason why we become fat when we eat a lot of food is as follows: Acetyl CoA[*5], an intermediary metabolite of glucose, is the starter for lipid biosynthesis, and overabsorbed glucose is stored in the adipose tissue, not as glycogen but as lipid owing to its efficiency as an energy stock.

On the other hand, higher organisms synthesize new compounds through an anabolic process to maintain biological activity, using high energy compounds biosynthesized through catabolism. Phosphoenolpyruvate, creatine phosphate, ATP, glucose-6-phosphate are cited as high energy compounds in the body, and these compounds discharge large amounts of free energy when they are hydrolyzed[*6]. ATP is one high energy compound among these that is most extensively used in the body, and creatine phosphate is used to maintain the ATP level when ATP is spent on muscular movement. It is established that creatine phosphate provides energy for several seconds at the beginning of a 100-meter dash.

As energy is consumed, the concentration of intracellular glucose decreases, and the blood glucose level also starts to drop in order to compensate for this. The brain, receiving this information through the circulation of blood, then acts on the liver through hormones to activate enzymes that break down stored glycogen to supplement glucose. When glycogen subsequently runs out, neutral lipid in adipose cells breaks down and free fatty acids are released into the blood. Free fatty acids are oxidized and contribute to ATP production as mentioned above. Meanwhile, it is also known that gluconeogenesis[*7] is accelerated using pyruvic acid and amino acids. In this manner, higher organisms maintain homeostasis to normalize their biological activity by considerably altering their metabolism to maintain ATP concentration in the body, making use of its production as an indicator. It is therefore recognized today that metabolism is controlled to maintain homeostasis through a complex interplay of information within the body.

Along with the conventional view that the production of substances and energy necessary for cell activity are controlled by the expression of genes and enzymes, a new idea in recent nutritional science research has emerged that metabolic changes can control gene expression. In order to understand cellular function, therefore, it has become crucial to measure all metabolites (metabolome), as well as the manifestation of genes and proteins, and to clarify the relationship between gene manifestation and enzyme activity. This method is referred to as "metabolome analysis,"[*8] and it enables us to carry out a more extensive analysis owing to advances in analytic technologies. The interdisciplinary application of these technologies is anticipated not only in the explication of biological phenomena but the identification of food-producing centers, the development of new drugs, and the diagnosis of disease.

科学解説

＊1 glucose［グルコース］

$C_6H_{12}O_6$ で表される炭水化物です。体の中では，解糖系，クエン酸回路により分解されて，そのとき発生するエネルギーを用い，高エネルギー化合物（ATP）を産生します。

＊2 glycogen［グリコーゲン］

グルコースがつながったもので，体の中では，グルコースの貯蔵型と考えてください。肝臓や筋肉にグルコースをグリコーゲンにする酵素が含まれており，食事から吸収されたグルコース濃度が血液中で高くならないように，筋肉や肝臓で積極的に取り込んでいるのです。筋肉も代謝調節臓器なのです。

＊3 neutral lipid［中性脂肪］

脂肪酸が3分子がグリセロール1分子に結合したもので，トリアシルグリセロールとも呼ばれます。植物油，バターなどの主な成分です。体の中に吸収された後に脂肪組織に蓄積し，エネルギー源として利用されています。グルコースが代謝されるときに生じるアセチルCoAから脂肪酸が生合成されるので，甘いものの摂り過ぎでも太るのです。

＊4 reduce, oxidize［還元する，酸化する］

酸素分子と結合する場合を「酸化」といい，「酸化された物がもとにもどる」ことを「還元」といいます。水素原子の動きからすれば，「水素原子を失う」反応が「酸化」であり，「水素原子を受け取る」反応が「還元」とも考えられるのです。

＊5 acetyl CoA［アセチルCoA］

補酵素A（CoA）と他の有機化合物（アセチル基などのアシル基）がチオエステル結合することで反応性の高い化合物となります。例えば，アセチル基と結合したアセチルCoAは糖質代謝や脂質代謝の中心的な化合物となります。

＊6 hydrolyze［加水分解する］

「水解」ともいわれ，化合物に水を作用させて，それを構成する2つの物質に分解する反応です。塩と酸を塩基に分解する反応，ペプチド結合が切断されてアミノ酸を生じる反応，中性脂肪から脂肪酸が生じる反応などがあります。

＊7 gluconeogenesis［糖新生機構］

脳などの一部の組織は必要なエネルギーを完全にグルコースだけに頼っており，哺乳類は，グルコースの分解と合成の両方を調節して血液中のグルコース濃度を一定に保っています。肝臓に蓄積していたグリコーゲンは約16時間ですべて消費されてしまいますので，血液のグルコース濃度維持には，他の有機化合物からグルコースを合成する必要があります。これを糖新生といいます。

＊8 metabolome analysis［メタボローム解析］

細胞の働きを包括的に理解しようと，DNAの情報から解析したものをゲノム解析，タンパク質の側から網羅的に解析したものをプロテオーム解析などと呼び，生化学，栄養学の研究分野で積極的に導入されてきましたが，近年，代謝産物は生命活動の最終の化

学的表現型で，この表現型を無視して生命活動や疾病を理解することはできないという考えになってきました。しかし，細胞内に存在する代謝産物は無限にあるといっても過言ではなく，そのため，予測不可能な多数の代謝産物を単時間で精度よく解析できませんでしたが，ここ数年，解析技術の進歩により，代謝産物の網羅的な解析が行われるようになり，これをメタボローム解析と呼びます。

英語 註

p.41

1	self replication：「自己複製」	
3	homeostasis：「恒常性」	
4	catabolism：p.35, 14 参照	
4	anabolism：p.35, 15 参照	
6	proliferation：「増殖」	
6	differentiation：「分化」	
7〜8	to be more precise：「もっと詳しく言うと」	
13	medium：「手段」複数系は media	
19	adipose：⇒ p.17, 19 参照	

p.42

9〜10	intermediary metabolite：「代謝中間体」	
10	biosynthesis：「生合成」	
15	biosynthesize：前項 biosynthesis の動詞形	
15	phosphoenolpyruvate：「ホスホエノールピルビン酸」	
15〜16	creatine phosphate：「ホスホクレアチン」	
16	ATP：⇒ p.38, 科学解説＊6 参照	
16	glucose-6-phosphate：「グルコース-6-リン酸」	
22	100-meter：「100 メートルの」	
23	consume：「消費する」	

p.43

1〜2	free fatty acid：⇒ p.38, 科学解説＊4 参照	
4	pyruvic acid：「ピルビン酸」	
12	expression：「発現」	
16	metabolite：⇒ p.28, 6 参照	
22	identification of food-producing center：「食品産地の特定」	

Metabo (Metabolic Syndrome) of "Food" and *Metabo* of "Information"

1   You are most likely to have given some thought to what implications the act of eating has on your body. In this chapter, we'll consider "food" and "nutrition" from the perspective of "information." These days metabolic syndrome[*1] has become a serious issue due to our easy access to high-cal-
5   orie foods. We are able to cope with an environment that fails to fully satisfy us, but we find ourselves surprisingly fragile when all our needs are satisfied ; this may apply to information as well as diet.

   Most of the food that we eat in our daily diet are essentially living organisms. For instance, fish and meat come from animals, and vegetables and
10 fruit originate from plants. (The sole exception is "salt," which is a mineral.) Thus there is no food that exists to be eaten. (The exception is mothers' milk.) Therefore, such organisms' own "information" is encoded in the food. Meat, for example, is originally muscle encoded with the information necessary for movement, while vegetables ought to possess the needed in-
15 formation to carry out photosynthesis. I think it is interesting to consider food as a "bearer of external information" in this way since it allows us to rethink the meaning of eating from a different perspective. We may have a yen for meat one day, salad another. We don't, by any means, however, hunger for the things encoded with information specific to an organism,
20 such as the mechanisms for movement called muscle (skeletal muscle) or

those for photosynthesis. What our bodies crave are the elements of these. Since these elements are no longer specific to that living thing, they, in a sense, become general "information," accessible to all. What is necessary for us is not the specific information of living things but such things as the amino acids or the glucose, which are the elements of the protein and the starch of particular organisms respectively.

A series of events, namely digestion, absorption, and metabolism takes place in our body to accomplish the "intake of information" referring to the act of eating. We could put it this way : Digestion is the process that deconstructs the specific biological information of the food, the so-called external information, to the general biological information ; absorption is the process that introduces the simplified information into the internal body ; metabolism is the process that reconstitutes and utilizes the absorbed information based on its own intrinsic information.

Now let us turn our attention to metabolic syndrome, commonly referred to in Japanese as "*metabo*." Metabolic syndrome is the state in which the accumulation of visceral fat, or what we call "obesity," causes insulin resistance, which not only affects the normal control of blood glucose, but also causes complications such as an abnormality in the metabolizing of lipids and hypertension that finally render the body prone to arteriosclerosis[*2]. The essential factor in metabolic syndrome is often explained as follows : Since the dawn of their time on earth, human beings have lived through almost every era of history under conditions where food was scarce, and the human body has evolved to cope with "starvation" at the genetic level (this is referred to as the "thrifty gene"[*3] hypothesis). As the human genetic makeup never took into account the state of nutritional abundance we now live in, our body lacks the internal mechanisms to cope with this. This is how "metabolic syndrome" has come about. In other words, this situation produces a paradoxical phenomenon in which the body, unable to cope adequately with an abundance of nutrients, ironically becomes unable to metabolize one of them, namely glucose, within the body.

Can't the same be said of "information"? Through the process of evolution humans have acquired "language," invented "characters," and efficiently shared and spread "information" regarding the external world. This can be referred to as "the process of civilization." Although "information" has qualitatively and quantitatively brought about great change, such change was sluggish throughout almost all the periods of civilization, allowing us to adequately adapt through our capacities. Until quite recently, periods of "information starvation" in which "information" was hard to come by lasted a long time. During these periods, humans actively sought and obtained the minuscule amounts of available "information," arranging and reintegrating these fragmentary pieces of information to cope with various matters. That is to say, we are accustomed to dealing with a shortage of "information." However, for the last ten years or so, "information" itself has undergone a revolutionarily change in both its quality and quantity with the spread of the Internet and portable communication terminals. The recent dramatic transformation in "information" has altered our process of assimilating it from an active to a passive pursuit. I have the feeling that we recipients of this torrent of "information" do not yet have the capacities to digest all of it. We seem to be facing, so to speak, a sort of "informational metabolic syndrome," don't we? In the midst of this abundance of information, I believe, we are unable to digest and integrate the information to a point that we are now, conversely, "starving for information." I would like to call this condition a state of "information resistance."

Metabo (Metabolic Syndrome) of "Food" and *Metabo* of "Information"

1   Surmounting the ill effects of abundance, as found in "metabolic syndrome" in the body or "informational metabolic syndrome," will present us with a major task. Exercise and improving dietary habits are advocated to mitigate "metabolic syndrome" within the body. So it may prove interesting to ask ourselves what will be required to cope with our "metabolic syndrome of information."

科学解説

＊1　metabolic syndrome［メタボリックシンドローム］

　　　　内臓脂肪の蓄積（いわゆる肥満）が原因で，インスリン抵抗性（インスリンの情報が体内で正確に伝わらなくなる状態）となり，糖尿病（高血糖），高脂血症，高血圧症が重複して発症し，動脈硬化症のリスクを高めているとする疾患概念のことをいいます。近年の，飽食と運動量の低下による生活習慣が原因であると考えられます。

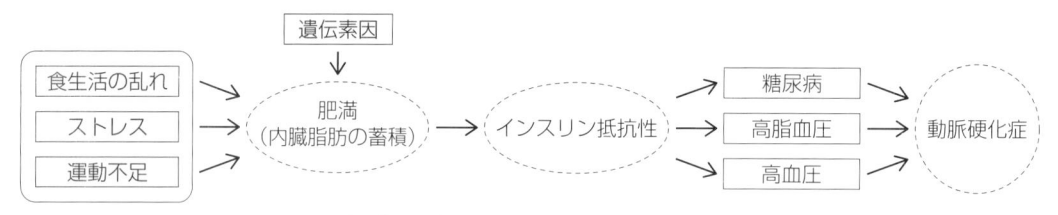

```
           遺伝素因
             ↓
 食生活の乱れ →
 ストレス   →  肥満        → インスリン抵抗性 → 糖尿病    →
 運動不足   → （内臓脂肪の蓄積）                 高脂血圧   → 動脈硬化症
                                                 高血圧    →
```

メタボリックシンドロームの原因と機序

＊2　arteriosclerosis［動脈硬化症］

　　　　動脈硬化症とは，動脈血管という管の壁が，コレステロールや中性脂肪などがお粥のような状態で固まって盛り上げられ，血液の流れが悪くなってしまった状態になり，最終的には血液が詰まってしまう疾患のことをいいます。脳血管で起こる動脈硬化症が脳梗塞であり，言語障害，半身麻痺，ひどい時には死に至ることもあります。また，心臓に血液を送る冠動脈で動脈硬化が起こると，心筋梗塞や狭心症になります。

＊3　thrifty gene［倹約遺伝子］

　　　　人間には，体内に効率よくエネルギーを蓄積できるようにして，飢餓状態における生存確率を上げるためのシステムが備わっており，それを担う遺伝子のことをいいます。倹約遺伝子という特定の遺伝子があるのではなく，エネルギー代謝にかかわる遺伝子に倹約型と標準型の２つが存在しているわけです。例えば，アドレナリンと反応して脂肪分解に関与しているβ3アドレナリン受容体遺伝子があげられます。日本人の40％が倹約遺伝子の頻度が高く，肥満となりやすいといわれています。

英語 註

p.46
18　　　yen：「熱望」
20　　　skeletal muscle：「骨格筋」

p.47
1　　　　those：＝ the mechanisms
2〜3　　in a sense：「ある意味で」
7　　　　series：⇒ p.21, 8 参照

7～8	take place：⇒ p.11, 10 参照	
17	visceral fat：「内蔵脂肪」	
17～18	insulin resistance：「インスリン抵抗性」	
19	complication：「合併症」	
20	hypertension：「高血圧」	
26	makeup：「構造」	
26	take into account：「考慮する」	

p.48

1	come about：「生じる，起こる」	
2	phenomenon：⇒ p.11, 6 参照	
9	bring about：⇒ p.30, 12 参照	
12	be hard to come by：「入手困難である」	
16	that is to say：⇒ p.9, 17 参照	
16	be accustomed to ～ ing：to の後は（動）名詞	
23	so to speak：「いわば」	
27	information resistance：「インフォメーション抵抗性」	

p.49

4	prove：⇒ p.17, 25 参照	

Too Much Is as Bad as Too Little —Diabetes[*1] Mellitus—

1   The increase of people suffering from diabetes has become a big issue in recent years. But just what kind of disease is this? You'd probably think that it is an affliction arising from an indulgence in a rich diet, wouldn't you? It certainly is true that the excessive intake of nutrition and a lack of
5 exercise can trigger diabetes. Yet if you were to peek inside a cell of a diabetic, you would find it resembles that of a starving person. Diabetes has the characteristics of a "pathogenic deficiency," rather than a "pathogenic abundance." Let's now consider what diabetes essentially signifies.

    Diabetes has existed from ancient times throughout the world. This afflic-
10 tion appears in historical records of 3500 years ago in ancient Egypt, referred to as an illness of "polyuria." It has recently become a national epidemic with one-sixth of the population said to be prediabetic in Japan. Fujiwara, Michinaga one of the most popular statesman of the Heian Period in Japan, suffered from the typical symptoms of diabetes : a thirst that drove
15 him to drink excessive amounts and feelings of listlessness. He is the first recognized person in Japan's recorded history to be afflicted with diabetes. So famous was he for this that he appeared on the commemorative postage stamp for the 15th International Diabetes Federation Congress held in Japan in 1997. The great inventor Thomas A. Edison also suffered from dia-
20 betes owing to his erratic schedule : He was absorbed in his research and inventing— cat-napping for a few hours, and eating lunch in the middle of night. You may think that such personages as to have their "high-living

diet have become diabetic from old times. Your image of diabetes is probably an affliction of extravagance, the body becoming, essentially, a big candy due to excessive intake of sugar. I wonder whether that is true. Let's consider the mechanics of the disease.

Diabetes is a disease in which the blood glucose levels increase ; specifically, it is caused by an anomaly in the mechanism that lowers the level of blood glucose. Glucose is the source of energy throughout our body. We commonly acquire glucose from sugar and starch contained in our food. This glucose is sent throughout the whole body via the blood, and its level in the blood is strictly controlled. When the level of glucose in the blood increases, the body lowers it by incorporating glucose into cells. When the blood glucose level drops, the body raises it by releasing glucose from organs such as the liver.

Now, let's closely examine the mechanism of lowering the blood glucose in detail. In order to lower the level of blood glucose, a "command" in the body is necessary. The hormone which bears this command is "insulin."[2] Insulin is like a ball imprinted with the command telling cells to absorb glucose from the blood. This ball is produced in the pancreas, and it is "thrown" into the blood. It then flows through the blood. Muscle and adipose cells have insulin receptors[3] on their surface, the "gloves" that catch the "insulin ball." Through this process, the receptor transmits the command to the cells. Finally, the cells open their gates to take in the glucose. The glucose that enters the cell is then used as a fuel to produce energy, and surplus glucose is stored as a form of fat called neutral lipids to stockpile energy.

Diabetes develops when the pancreas stops producing insulin, when insulin is produced but the receptors on the cell membrane of the muscle and adipose cells fail to function properly, or when the transmission of the command to the cell interior is disrupted. Research into these mechanisms continues to progress, revealing that the defect of insulin production largely depends on genetic factors. It is caused by a condition in which the β cells of islets of Langerhans[*4] in the pancreas (which is basically like a factory complex that produces insulin) are decimated by body's "defense forces" — the immune system—running amok. Various studies are progressing on other reasons for abnormality in insulin receptors and the transmission of the command. Recently, however, it has become clear that obesity is one of the major causes of diabetes. It is said that the increase of fat intake due to a Westernized diet and a lack of exercise have led to an increase in body fat, which, in turn, hinders the transmission of the command found in insulin. Further study is required to clarify the intracellular mechanism of diabetes.

As mentioned above, this defect in the transmission of the command in insulin impairs the absorption of glucose into cells, finally bringing about an increase in blood glucose levels. However, considering this situation on a cellular level, intracellular glucose levels actually drop, causing an energy deficit. Since cells cannot survive in this state, they try to produce energy by decomposing fat. This, consequently, generates ketone bodies.[*5] The increase in ketone bodies makes our body acidic, producing a state referred to as ketoacidosis.[*6] Ketoacidosis, in fact, also occurs in a state of starvation. Since there is a lack of glucose in a state of starvation, the intracellular glucose level also decreases. Although diabetes and a state of starvation would seem to be contrary to each other, they share the same type of intracellular impairment. How ironic it is that a condition of deficit and one of excess both result in the same impairment! As Confucius said, "Too much is just as bad as too little," and this certainly rings true in the course of our lives.

科学解説

*1　diabetes（diabetes mellitus）［糖尿病］

　　　　血液中のブドウ糖（グルコース）濃度（血糖）が，病的に高くなる疾患のことをいいます。症状は多岐にわたり，無症状のものもあれば，のどの渇きや大量の尿を出すもの，重篤になると意識障害を起こし，手足の壊疽（えそ），腎障害をひき起こすものもあります。この病気の原因は，血糖値を調節するホルモンであるインスリン（次項で解説）の機能障害です。インスリンの機能障害にも2種類あり，インスリン合成・分泌細胞である膵臓ランゲルハンス島β細胞が障害された結果インスリン分泌障害が起こすⅠ型と，生活習慣などによりインスリン分泌低下とインスリン受容体によるインスリン感受性の低下（インスリン抵抗性という）によって起こるⅡ型が存在します。

*2　insulin［インスリン］

　　　　インスリンは，生体内で血糖降下作用を有する唯一のホルモンです。21個のアミノ酸が結合したA鎖と，30個のアミノ酸が結合したB鎖で構成されるペプチドホルモンです。膵臓のランゲルハンス島β細胞で合成されます。インスリンの主な標的臓器は，肝臓，筋肉，脂肪組織です。インスリンは，これらの標的細胞に血液中のグルコースを取り込み，利用を促進させる機能をもっています。

*3　insulin receptor［インスリン受容体］

　　　　インスリン受容体は，インスリンの標的臓器の細胞表面に存在している膜貫通型タンパク質です。インスリンがインスリン受容体に結合すると，細胞内にインスリンの情報が伝達されていきます。

*4　islets of Langerhans［ランゲルハンス島］

　　　　膵臓は，胃の後方にある赤みがかった灰白色の長さ約15 cmの臓器です。膵臓には，大きく2つの機能があり，ひとつは消化酵素を産生・分泌する外分泌機能，もうひとつは，血糖調節を担うホルモンを産生・分泌する内分泌機能です。膵臓は，90%以上が外分泌細胞である腺房細胞です。残りの数%がランゲルハンス島と呼ばれる内分泌細胞の塊です。この部分のβ細胞で血糖値を低下させる唯一のホルモンであるインスリン，α細胞で血糖値を上昇させる役割をするグルカゴンが分泌されます。

*5　ketone body［ケトン体］

　　　　絶食状態など食事からの糖質の供給が途絶えた場合，体内に蓄積されている脂肪酸がエネルギー源として動員されます。そのとき，肝臓では，脂肪酸がβ酸化という過程でアセチルCoAを作り出し，TCAサイクルで処理されることでエネルギーが作られています。しかし，脂肪酸が大量の場合，アセチルCoAがTCAサイクルで処理しきれず，アセト酢酸やβ-ヒドロキシ酪酸を作り出し，またアセト酢酸はアセトンになります。これら，アセト酢酸，β-ヒドロキシ酪酸，アセトンの3種の化合物をケトン体といいます。

*6　ketoacidosis［ケトアシドーシス］

　　　　ケトン体は，肝臓以外の組織ではエネルギー源として利用できます。絶食状態が長期

に継続すると，脂肪酸を動員しエネルギー供給に使わなければなりませんが，血中ケトン体の増加によってケトーシスが起こります。また，ケトン体のアセト酢酸，β-ヒドロキシ酪酸が体内で増加すると，体内が酸性となり（pHが下がり）アシドーシスという状態になりますと重篤になると死に至ります。糖尿病でもケトアシドーシスを呈することがあります。

英語 註

p.52

タイトル　too much is as bad as too little：「過ぎたるはなお及ばざるがごとし」（ことわざ）

タイトル　diabetes：単数扱い

5　　if you were to～：「もし～するとしたら」（ほとんど可能性のない強い仮定を表す）

6　　that：＝ the cell

7　　pathogenic：⇒ p.29, 13 参照

11　　polyuria：「多尿症，頻尿症」

12～13　Fujiwara Michinaga：「藤原道長」(966-1027 日本の平安時代中期の政治家)

17　　so famous was he：倒置

18　　Internaional Diabetes Federation Congress：「世界糖尿病学会」

22　　high-living：「ぜいたくな生活の」

p.53

6　　anomaly：「異常」

18　　pancreas：⇒ p.36, 1 参照

19～20　adipose：⇒ p.17, 19 参照

24　　neutral lipid：「中性脂肪」

27　　cell membrane：⇒ p.36, 4～5 参照

p.54

5～6　complex：「複合体」

7　　run amok：「逆上して暴れる」

15　　bring about：⇒ p.30, 12 参照

25　　one：＝ a condition

26　　Confucius：「孔子」(551?-479B.C. 中国春秋時代の思想家，儒教の開祖)

27　　ring～：「～のように聞こえる」

Cancer is Complex

1   According to the Ministry of Health, Labor and Welfare, "cancer" has been the primary cause of death among the Japanese people since the 1980s, and it currently accounts for about 30% of all causes of death. Today, cancer is considered as a "chronic disease arising from the accumula-
5   tion of the genetic abnormalities." Among such genes relevant to the development of cancer are those belonging to the oncogene family, which accelerate the spread of cancer. An oncogene is a proto-oncogene mutant[*1], and these proto-oncogenes normally manifest themselves in normal cells, contributing to the maintenance of normal cell functions. So it is throught that
10 the loss of normal function of the proto-oncogene with a mutation disturbs intracellular order and exerts its influence further. It may be easier to understand this if you imagine something like a rotten apple in a box that spoils the others one by one. All proto-oncogenes share the function of encoding proteins related to the "signal transduction pathway." It is consid-
15 ered that a cell will become cancerous when it mutates under influence of external stress and aging or loses its function because of a heterozygous defect[*2].

   On the other hand, an anti-oncogene (tumor-suppressor gene) is something like a brake that protects a cell from canceration if you liken an onco-
20 gene to a car's accelerator. Anti-oncogenes neither manifest themselves normally nor do they exert any influence in most hereditary familial tu-

mors[*3].

These oncogenes and anti-oncogenes control some common processes of cellular activities. I would now like to turn to some concrete examples to describe what is known about their affective mechanisms.

For instance, the division and proliferation of normal cells are strictly controlled and regulated by factors involved in the cell cycle[*4]. A cell cycle generally consists of the two processes of DNA replication and cell division, and it is controlled by extracellular signals—it is believed, however, the cell cycle in cancer cells runs amok to cause misevolution. Tumor-suppressor proteins such as P53 and Rb are cited as factors that control cell cycles. P53 and Rb have the role of stopping a cell cycle when it fails to function normally. Numerous studies have reported cancer cells in which the impairment of p53 has been found. To return to my previous metaphor, imagine a car that has no brakes. Factors that relay extracellular signals into the cell and invigorate cell proliferation, on the other hand, have been also recognized : Ras protein is activated by PDGF and EGF, hormones that actively stimulate the processes of cell proliferation and differentiation. Once activated, Ras contributes to cell proliferation and differentiation by transmitting its information to the nucleus and activating a gene manifestation mechanism in the right time. Of importance is the subsequent presence of a mechanism that suppresses unnecessary cell proliferation and differentiation through a process of deactivation. Ras was discovered as an oncogene of the sarcoma virus in rats around 1980, and it was found that signals to proliferate were continuously being transmitted to the nucleus since the Ras, as a mutation, remained in an invariably active state within the cell. In other words, because the gene that encodes Ras protein in cancer cells undergoes a mutation, and this Ras protein remains constantly active within the cell, it cannot be controlled by hormones . You could say this is like a car with a driver who never lets up on the accelerator.

The next cell function I would like to consider is apoptosis. Apoptosis is a function inherent in a cell, and it is the "program" by which a cell terminates itself, so to speak, a programmed cell death mechanism. For example, the mammalian "hand" is initially webbed but through apoptosis the cells of this webbed area are shed during embryonic development to form fingers. It is further postulated that apoptosis is involved in the decrease of lymphocytes from AIDS and the dementia associated with aging. A more typical example would be when skin peels after a sunburn. We now also know that the health of our body is maintained by shedding cancerous cells through their removal by apoptosis. In other words, a cancer cell must terminate the apoptotic mechanism in order to survive in our body. The P53 protein, a tumor-suppressor protein previously mentioned herein, inducs apoptosis when DNA anomalies occur during the course of the cell cycle.

Recent studies have additionally come to reveal that the manifestation of oncogenes and anti-oncogenes change not simply through the mutation of the genes themselves but also through "epigenetic"[*5] modification. That is to say, we are gradually coming to understand from a variety of viewpoints how the obuscure stressors of daily living and aging are linked to the onset of cancer.

In this manner we have come to understand the functions of oncoproteins and tumor-suppressor proteins little by little, but a thorough method for the treatment of cancer remains elusive to this day. This is due to the complicated mechanisms involved in the onset of cancer and the fact that they all do not converge into one simple process.

> 科学解説

*1 mutant［変異体］

　　　遺伝子（DNA）は細胞の核の中に格納されており，生命活動を行うのに必要なタンパク質の構造を決定する情報を含んでいます。すなわち，タンパク質が建築物ならDNAは設計図であり，核は設計事務所といったところです。DNAはアデニン（A），グアニン（G），シトシン（C），チミン（T）のそれぞれ異なる塩基をもつヌクレオチドが結合し合い長い鎖を作っています。A，G，C，Tの並び方によりタンパク質の構造に関する情報が暗号化されています。例えば，AGCTTTATCGTATTGTAGTAGGといった感じです。ところが，年齢を重ねたり，発癌性物質が体内に取り込まれたり，外部環境のストレスの影響によりこの配列に変異が生じてしまうことがあります。先程の遺伝子配列がAGCTTTATCTTATTGTAGTAGGに変わってしまい，つまり，設計図にミスが生じ，手抜き工事の建造物ができてしまったようなものです。これを，変異体といいます。大事なことは，変異が構造に大きな影響を与えるところに生じてしまった場合にのみ，重大な欠陥となることです。

*2 heterozygous defect［ヘテロ接合性欠損］

　　　私たちの遺伝子は，父親由来のものと，母親由来のものと2本存在しています。癌抑制遺伝子の場合，父親由来の遺伝子に変異があったとしても，母親由来の遺伝子が正常であれば，細胞機能は正常に機能します。しかし，何らかの理由により，生まれた後に母親由来の癌抑制遺伝子にも変異が生じ，父親，母親のどちらの遺伝子からも癌抑制タンパク質が作られなくなります。この現象をヘテロ接合性欠損といいます。

*3 familial tumor［家族性腫瘍］

　　　遺伝性の癌を家族性腫瘍といいます。母親もしくは父親が変異遺伝子のキャリアで，ヘテロ接合性欠損により癌を発症します。

*4 cell cycle［細胞周期］

　　　ひとつの細胞は決まった順序で起こる一連の過程によって倍加し，2分して増えます。この一連の過程を細胞周期といいます。受精した直後の細胞は1個なのに，成人になると約60兆個以上になるといわれています。すなわち，1個の細胞が細胞周期を繰り返すことで60兆個にまで増えるのです。私たちの体の細胞は幹細胞などの一部を除き基本的には細胞周期は静止期にあります。

　　　細胞周期は核の分裂をともなう有糸分裂と，細胞が2つに分裂する細胞質分裂とにわかれます。有糸分裂ではDNA量が倍加し，細胞が2つに分裂したときに均等に遺伝子情報をそれぞれの細胞に分けます。

*5 epigenetic［エピジェネティク］

　　　形質発現は遺伝子の発現により規定されていると考えられてきました。すなわち表にでる表現系（花が赤いとか，髪の毛が黒いとか）は遺伝子の配列で決まっていると考えられてきました。ところが最近の研究では，遺伝子を核の中に格納しているクロマチン

タンパク質やDNAの化学修飾が遺伝子の発現に影響を及ぼしていることが解ってきました。例えば,「…を食べて体質が変わった」などはその例として考えられています。また,食生活が大きな要因を占める生活習慣病の発病にも大きくかかわっていることが解ってきました。

英語 註

p.57

1	the Ministry of Health, Labor and Welfare：「厚生労働省」
3	account for ～：「～を占める」
5～6	among such genes relevant to the development of cancer are those：倒置
6	those：＝the genes
6	oncogene：「がん遺伝子」
7	proto-oncogene：「がん原遺伝子」
12	apple：日本ではみかんの譬（たとえ）話だが,アメリカではりんごを使う。
18	tumor-suppressor gene：「がん抑制遺伝子」

p.58

5	proliferation：⇒ p.41, 6 参照
7	replication：「複製」
8	extracellular：「細胞外の」
9	run amok：⇒ p.54, 7 参照
11	it：＝a cell cycle
16	PDGF：「血小板由来増殖因子」
16	EGF：「表皮増殖因子」
17	differentiation：⇒ p.41, 6 参照
20	of importance：＝ important
23	sarcoma virus：「肉腫ウイルス」

p.59

1～2	a car with a driver who never lets up on the accelerator：「運転者がアクセルを踏みっ放しの車」
3	apoptosis：「アポトーシス,プログラムされた細胞死」
5	so to speak：「いわば」
6	mammalian：「哺乳類の」
8～9	lymphocyte：「リンパ球」
15	anomaly：⇒ p.53, 6 参照
18～19	that is to say：⇒ p.9, 17 参照
22	oncoprotein：「がんタンパク質」

Researching the Global Environment at the Bottom of the Earth

1   Global warming, holes in the ozone, pollution from man-made chemicals —human activity is drastically altering the global environment. In particular, global warming caused by the greenhouse effect[*1] is producing such menacing conditions as climate change, vanishing territories due to rising
5   sea levels, and the spread of infectious tropical diseases. While various observations have been made in many different nations and regions, investigations and research relevant to global warming are also going on in the frigid region of Antarctica at the bottom of the earth. Here, I would like to present some of the research carried out in Antarctica, an area far removed
10 not only from Japan but also from other nations where human beings are active.

   The amount of greenhouse gases[*1] in the atmosphere, especially that of carbon dioxide (CO_2), has become a problem as the main cause of global warming. But only 50 years have passed since human beings first began
15 the periodic measurement of carbon dioxide. As atmospheric composition varies over time, it must be measured at particular instances, but it is impossible to go back in time and conduct measurements. If only we had a time machine, it would be a different situation, however…; in fact, we do have a time machine. Where, you ask? Inside the thick layer of ice covering
20 the center of Antarctica. And it is not a machine made by aliens that you'd see in some second-rate SF flick but the very ice itself. The altitude of the

Antarctic Continent is low and its interior is covered with an ice sheet[*2] which has a maximum thickness of 4,776 m. Since the temperature is extremely low in the inland region of Antarctica, the snow lying on the surface of this ice sheet never melts in any season. As a result, new snow keeps accumulating on top of the old snow, pressing and hardening it into ice with air trapped within its pockets.

Therefore, the ice near the surface of the ice sheet contains the atmospheric components of a relatively recent past, while the ice in deeper layers contains those of a more distant past. Time-sequential changes in atmospheric components from the past to present can then be analyzed in the following manner : first, the ice layer is drilled vertically at a given point to extract a column-shaped ice, which is referred to as an "ice core" ; the ice core is then cut into short round slices and analyzed to discover the atmospheric components of each slice ; and finally, putting the slices in a time-sequential order enables researchers to analyze the changes in atmospheric components from past to present. This is the above-mentioned time machine. At Dome Fuji Base located in the eastern inlands of Antarctica at an altitude of 3,810 m, the Japanese Antarctic Research Expedition started drilling in 1995 and has so far extracted an ice core with a maximum length of 3,035 m.

Analysis of this core threw light on climate changes, including atmospheric components and temperatures, over the past 340,000 years. The results indicate that a glacial period followed by an interglacial period reoccurred 3 times in the course of 100,000 years over the past 340,000 years, and for roughly the past 10,000 years the earth has been in an interglacial period and is gradually cooling. In addition, compared with the carbon dioxide levels prior to the Industrial Revolution which were roughly 300 ppm at most, it became clear that the current concentration of about 375 ppm is extremely high. In spite of the earth progressing into a mild cooling period, temperatures continue to rise and the average temperature of the earth has increased by 0.74℃ over the past century and by approximately 0.3℃ from 1990 to 2007. IPCC (Intergovernmental Panel on Climate Change)[*3] estimates that, compared to that of the year 2000, the average temperature may increase by as much as 6.4℃ by the end of this century if the carbon dioxide levels in the atmosphere continue rising at the present pace.

I will change the topic—numerous micro-particles called "aerosols,"[*4] in other words "dust," float in the air. And certain kinds of aerosols form the cloud condensation nuclei that constitute clouds. Since aerosols and clouds reflect a portion of the sunlight in the atmosphere, they reduce the quantity of the sunlight that reaches the surface of the earth and suppress the rise in temperature. Aerosols are released in large quantities through human activity, but they are also produced by marine organisms. We smell the sea when we go to the beach, and the origin of this smell is a gas called dimethyl sulfide (DMS)[*5], which is emitted from seaweed and phytoplankton[*6]. It is now understood that after being released into the atmosphere, DMS is oxidized and becomes small sulfate particles, and these particles rise into the sky to form the nuclei of clouds.

An innumerable number of Antarctic krill[*7], zoo plankton that resemble small shrimp, inhabit the Antarctic Ocean. These are referred to as a "key species"[*8] in the food chain[*9] of the Antarctic Ocean as they are the most important food source of whales, seals, penguins, and fishes residing there. Recent research has shown that the majority of Antarctic krill are distributed in areas of the sea that freeze over in winter (pack ice area[*10]) while zooplankton called "salpa"[*11] are distributed in open water. Both krill and salpas consume phytoplankton as their food, but Antarctic krill eat phytoplankton through mastication while salpas swallow them whole and digest them. This difference in eating is important : as mastication breaks down phytoplankton, it releases DMSP (a precursor of DMS) into the seawater, whereas swallowing does not break them down and consequently produces no DMSP. The amount of DMS released into the atmosphere, in other words, varies according to the way of consumption of phytoplankton.

Let's synopsize how the methods of phytoplankton consumption affect the environment : When Antarctic krill consume phytoplankton through mastication, the concentration of DMS in the atmosphere increases and clouds form more easily to reflect sunlight and decrease atmospheric temperatures ; conversely, when salpas swallow phytoplankton, clouds that reflect sunlight and lower temperatures fail to form since there is no increase in the concentration of DMS in the atmosphere. Because of the reduction of ice cover in the Antarctic Ocean in recent years as a result of global warming, the habitat of the Antarctic krill is contracting and that of salpas is expanding. And the Antarctic climate may possibly change should the population of Antarctic krill decrease. The Antarctic climate affects the circulation of the ocean waters as well as the atmospheric flow of the entire earth. In short, the Antarctic climate changes caused by global warming could alter the climate on a global scale.

Now scientists are locked in a heated battle to clarify the mechanism of changes in the global environment in the bitter cold of Antarctica at the bottom of the earth.

科学解説

＊1 greenhouse effect［温室効果］・greenhouse gas［温室効果ガス］

　　　　地球の気温は，太陽からの放射（太陽光）と地表からの赤外線による放熱によってバランスがとられています。この放熱の一部を地表面にとどめているのが二酸化炭素（CO_2）やメタン（CH_4），一酸化二窒素（N_2O）などの温室効果ガスです。地球の平均気温は約15℃です。これは温室効果ガスが宇宙空間に逃げていく赤外線を地球に戻しているからです。もし温室効果がないとすると地球の平均気温はマイナス18℃になると試算されています。もちろん，温室効果ガスが増加すれば地表面に戻る赤外線の量も多くなるので，気温も上昇します。近年，人類の活動，特に化石燃料の燃焼の増大にともなって大気中の二酸化炭素濃度も増加しています。人類の活動による温室効果ガスの急激な増加が，温室効果に拍車をかけて，地球の気温を急速に上昇させている原因であるというのが「地球温暖化現象」です。

＊2 ice sheet［氷床］

　　　　大陸規模の広大な地面を覆う巨大な氷の塊のことをいいます。氷床が長期にわたって存在するためには，降り積もった雪が解けないほど寒冷であること，地面があって雪を保存できることという2つの条件を満たすことが必要です。現在の地球上には，南極大陸上の南極氷床とグリーンランド氷床の2つがあります。

＊3 IPCC（Intergovernmental Panel on Climate Change）［気候変動に関する政府間パネル］

　　　　気候変動の原因や影響について最新の科学的，技術的，社会・経済的な知見を集約し，評価や助言を行っている国際機関です。IPCCが発表する報告書や数値資料などは，温室効果ガスの削減目標を定めた京都議定書の基礎になるなど，国際的に重視されています。

＊4 aerosols［エアロゾル］

　　　　日中晴れていても，空全体が白く霞んでいることがあります。これは大気中に浮遊する微粒子が，日射を散乱させているためです。気体中に微粒子が浮遊している状態を「エアロゾル」といい，気体中の浮遊している花粉や土ぼこりなどの大きさ数nmから数μm程度の微粒子を「エアロゾル粒子」といいます。エアロゾル粒子は日射の散乱・吸収物質として機能するだけでなく，雲ができるときの雲核（うんかく）として機能するため，気象に大きく影響しています。

＊5 dimethyl sulfide（DMS）［硫化ジメチル］

　　　　硫化ジメチル（ジメチルサルファイド：CH_3SCH_3）が大気中に放出されると，光化学反応により，エアロゾル粒子成分の前駆体である二酸化硫黄（SO_2），硫酸ガス，メタンスルホン酸ガスに酸化され，最終的には新粒子生成・成長・とりこみ過程を経てエアロゾル粒子になります。

＊6 phytoplankton［植物プランクトン］

　　　　海洋・湖沼・河川などの水中を漂いながら生活する単細胞または群体性（単細胞生物が連結された集合体）の藻類や微小な高等植物の総称です（大きさは0.2～200 μm）。植物プランクトンの中でもケイ酸質の殻をもった珪藻類は，生物量が著しく多く，水中の一次生産者として，水中の食物連鎖において重要な役割をしています。ちなみに，オキアミやミジンコ，クラゲなどほとんど遊泳力のない動物を動物プランクトン（zooplankton）と呼びます。

＊7　Antarctic krill［ナンキョクオキアミ］

　　　オキアミ類はエビに似た動物プランクトンで，全世界の海洋から約85種が報告されています。主に外洋域を生息場所とし，南極海には7種のオキアミ類が分布していますが，これらの中で最も個体数が多いのがナンキョクオキアミ（学名：*Euphausia superba*）です。ナンキョクオキアミは，南極海産のオキアミ類の最大種で全長6cmに達し，寿命は5～7年と推定されています。非常に大きな群れを作り，ときには1m^{3}に3万個体もの高密度で群れることがあります。ナンキョクオキアミは南極海で最大の生物量があり，その量は少なく見積もっても約2億tで，10億tを下らないという説もあります。主に植物プランクトンやアイスアルジー（氷下面に付着する珪藻類）を餌とし，一方ではクジラ類，アザラシ類，ペンギン類，魚類など，南極海に生息するさまざま動物の餌になっています。そのためナンキョクオキアミは南極海生態系の「鍵種」と呼ばれています。

＊8　key species［鍵種］

　　　莫大な生物量があることなどで，食物連鎖などにおいてさまざまな生物に利用され，その生態系が成り立つ上で重要な役割を担う生物種を「鍵種」といいます。

＊9　food chain［食物連鎖］

　　　体外から取り入れた有機物に栄養源を依存している栄養形式を従属栄養といい，栄養源を無機物だけに依存している栄養形式を独立栄養といいます。独立栄養を営み，無機物から有機物を合成する生物を「生産者」とよび，緑色植物や植物プランクトン，一部の光合成細菌などの光合成生物，あるいは硝化細菌や硫黄酸化細菌などの化学合成細菌が含まれます。従属栄養を営む動物などは，生産者が合成した有機物を摂取して栄養源にしているので「消費者」とよばれます。また，生産者や消費者の死骸や排出物に含まれる有機物を二酸化炭素や水，アンモニアなどに分解して環境に戻す生物を「分解者」といいます。生産者が作りだした有機物を第一次消費者が捕食することによって獲得し，次に第一次消費者を第二次消費者が捕食することによって獲得する，この「食う―食われる」関係（捕食被食関係）が鎖のように繋がっているので，食物連鎖と呼ばれています。

＊10　pack ice area［冬に氷の張る海域（流氷域）］

　　　氷床とは異なって，寒気にさらされた海面が凍ることによってできた氷を海氷といい，海氷はその運動の状態によって，定着氷（fast ice）と流氷（pack ice）に分けられます。定着氷とは海岸や氷床の縁などに接して水平方向へ動かない海氷のことで，流氷とは流れて動いているか動くことが可能な状態にある海氷のことです。南極海の海氷面積は，9月に最大（約2000万km^2）になり，2月には最小（約500万km^2）になります。このように南極海の海氷には季節による消長があることから，大部分が流氷であることが分かります。

＊11　salpa［サルパ］

　　　半透明な樽型をしている動物プランクトン。体が透明なため，クラゲのような印象を受けますが脊索動物門のホヤの仲間です。全世界の海洋に分布しますが，南極海で多い種の大きさは体長10cm程度。海の中を漂って生活し，海中の植物プランクトンを濾し取って餌にしています。近年，個体数の増加傾向が報告されていて，ナンキョクオキアミとの競合関係が注目されています。

英語 註

p.62

12	that：＝the amount	
17	if only ～：「～でありさえすれば」	
18	do：have を強調	
21	flick：「映画」	
21	very：⇒ p.22, 18 参照	

p.63

6	with air trapped within its pockets：「すき間に大気を閉じ込めたまま」	
6	it：＝the old snow	
9	those：＝the atmospheric components	
9	time-sequential：「時系列的な」	
17	Dome Fuji Base：「ドームふじ基地」	
18	the Japanese Antarctic Research Expedition：「日本南極地域観測隊」	
19	drill：「(穴を) ボーリングしてあける」	
19	so far：⇒ p.10, 7 参照	
27	at most：「多くても」	

p.64

4	by ～：「～だけ，～の差で」	
6	that：＝the average temperature	
7	as much as ～：「(数詞を伴い)～ほども」	
23	species：⇒ p.22, 18 参照	
27	open water：「氷が張っていない海面」	

p.65

2	mastication：「咀嚼（そしゃく）」	
4	DMSP：dimethyl sulfoniopropionate「ジメチルスルホニオプロピオナート」	
4	precursor：「前駆物質，前駆体」	
7	consumption：「摂取」	
9	consume：⇒ p.36, 21 参照	
16	contract：「縮まる」	
17〜18	should the population of Antarctic krill decrease：＝if the population of Antarctic krill should decrease「万一～ならば」	

Bioethanol

1   Global warming*1 is one of the environmental problems confronting the human race. Bioethanol*2 is attracting attention as a means toward solving this problem. This fuel alternative to gasoline is produced from biomasses derived from renewable living organisms such as plants. It is produced
5   through various advanced technologies employing microbes. In this article, I would like to introduce some cutting-edge technologies employed in the production of bioethanol and some of the problems inherent with this fuel.

   One of the many environmental problems that human beings currently face is that of global warming. This indicates a rise in the average tempera-
10 ture of the earth due to the reduction of forest area by exploitation or fire, the use of fossil fuels*3 and the generation of various greenhouse gases. Global warming triggers various problems, such as the submersion of low lands below sea level, fluctuations in available arable land, and a diminution of cultivated land. Various countermeasures toward these problems are
15 therefore being devised for sustainable human development. Here I would like to talk about one of these countermeasures : bioethanol.

   To begin with, do you know that the burning of plants, such as firewood, does not increase greenhouse gases? A plant grows by trapping carbon dioxide from the atmosphere through photosynthesis. Hence the carbon diox-
20 ide generated when that plant is burned is thought to be what was origi-

nally in the atmosphere. So in this case, although the burning of vegetation produces carbon dioxide as the greenhouse gas, it does not increase the overall amount of greenhouse gases in the atmosphere. This is the concept referred to as "carbon neutral."[4] On the other hand, fossil fuels are not substances formed through trapping initially airborne carbon dioxide. They are found underground in the form of solids, liquids or gas. The overall volume of carbon dioxide in the atmosphere consequently increases by burning fossil fuels since this releases amounts of the gas that were not originally there. This property is referred to as "carbon negative,"[5] the opposite of "carbon neutral." If we were therefore able to utilize carbon neutral energy, we could perform various activities without accelerating global warming.

Bioethanol is one type of such energy. It is ethanol produced through the exploitation of biomass, a resource derived not from fossil fuels but from renewable living organisms such as plants. And it is attracting attention as a biomass fuel to become an alternative to gasoline. It is produced through the utilization of various microbes. Myriad species of microbes, including those which as yet remain undiscovered, exist with diverse characteristics. One example of these would be what take advantage of sugar (glucose) and produce various substances during their growing process. Bioethanol is one of these substances, and its production employs a species of microbe that grows by making use of glucose to produce ethanol in a highly efficient manner.

In the United States attention was given to corn, rich in starch, as a biomass that these microbes could utilize. This type of starch, however, had to be physicochemically converted to sugar of which the microbes make use through the use of fossil fuels. A number of research institutes, therefore, examined various methods to solve this problem. One of these was a method devised to break down starch using an amylolytic enzyme from a certain species of microbe. The use of this enzyme allowed starch to be converted into a state that the microbes could take advantage of without the use of fossil fuels. Further, the gene of the microbe that produces ethanol with such high efficiency was modified with the gene of this enzyme to enable it to break down starch. These modifications allowed the microbe to both break down starch into sugar and convert the sugar into ethanol simultaneously. And in this way, the production of ethanol directly from starch became possible. The development of these various technologies has now made the production of bioethanol more efficient than it once was.

Since there are still various areas for improvement (such as the need for greater efficiency in both production of bioethanol and use of its material), however, the pursuit of further progress in these technologies continues. In Brazil, biomass ethanol is being produced from sugar cane. This bioethanol is already being mixed with gasoline or other fuels as well as being used as a fuel in itself. But the production of this type of bioethanol from corn and sugar cane brought about a decline in their supply as food materials and a steep rise in their price. Alternative means for obtaining the raw material of ethanol by breaking down cellulose, the major component of general plants, are therefore garnering attention.

It is difficult to convert all fuels to bioethanol soon considering the efficiency of the conversion from biomass to bioethanol. Yet advances in such technologies are indispensable for sustainable human development. And it just may be you, learning about this now, and no one else, who advances these technologies.

> 科学解説

*1　global warming［地球温暖化］

　　　地球温暖化とは，地球表面の大気や海洋の平均温度が長期的にみて上昇する現象です。単に「温暖化」ということもあります。地球の歴史上では，平均気温の上昇や下降が幾度となく繰り返されてきたと考えられています。そのため，「温暖化」は単に地球全体の気候が温暖に変わる現象を指すこともあります。しかし，ここでは，近年観測され将来的にも続くと予想される「20世紀後半からの温暖化」を指しています。この地球温暖化は，大気や海洋の平均温度の上昇だけではなく，生物圏内の生態系の変化や海水面上昇による海岸線の浸食といった，気温上昇に伴う二次的な諸問題をひき起こします。

*2　bioethanol［バイオエタノール］

　　　バイオエタノールは環境問題，原油依存脱却の点から現在注目されている新エネルギーです。サトウキビや廃木材，大麦やとうもろこしなどの植物資源を原料として生産されるエタノールのことです。バイオエタノールは一般的にガソリンと混合して，自動車やボイラー等の燃料として利用されます。日本では現在，法律で3％まで混合できることになっています。この混合燃料は，エタノールを利用する分，原油の消費が減り，二酸化炭素の排出量を削減し，地球温暖化防止に貢献します。さらに，廃棄物の削減にもつながる地球環境に配慮したエネルギーとして，大きく期待されています。

*3　fossil fuel［化石燃料］

　　　化石燃料は，人間の経済活動で用いられる，または今後用いられることが検討されている，有限性の燃料資源の総称です。地質時代に動植物などの死骸が地中に堆積し，長い年月をかけて地圧・地熱などにより変成されてできた有機物の化石です。主に利用されているものに石油や石炭，天然ガスなどがあります。また，近年ではメタンハイドレートなどの利用も検討され始めています。現在，人間活動に必要なエネルギーの約85％は石油から得ています。石油は，輸送や貯蔵が容易であることや，大量のエネルギーが取り出せることなどから使用量が急増しています。

*4　carbon neutral［カーボンニュートラル］

　　　カーボンニュートラルは環境科学の用語のひとつです。直訳すればカーボンは炭素，ニュートラルは中立なので「環境中の炭素循環量に対して中立」という意味になります。これは何らかの物質を生産する際や一連の人為的活動を行った際に，排出される二酸化炭素と吸収される二酸化炭素が等しくなることを表す用語です。例えば，植林をすると，植林した樹木が成長過程で光合成により，二酸化炭素（カーボン）を吸収することで空気中の二酸化炭素が減少します。そして，成長した木を伐採し，燃料として燃やすことで空気中の二酸化炭素が増加します。このサイクルでは，二酸化炭素の増減がプラスマイナス0となるため，カーボンニュートラルであるといえます。

＊5　carbon negative［カーボンネガティブ］

　　　　カーボンネガティブは環境科学の用語のひとつです。これは何らかの物質を生産する際や一連の人為的活動を行った際に排出される二酸化炭素が，吸収される二酸化炭素を上回る場合を表しています。また，日本においては「カーボンマイナス」ともいわれます。例えば，化石燃料は非常に長い時間をかけて地中に蓄積されます。これにより，過去の空気中の二酸化炭素が減少しました。しかし，また地中に蓄積するためにも非常に長い時間をかけるため，現在を基準にみると，空気中の二酸化炭素を増加させていると言えます。

英語 註

p.69
2　　　means：「手段，方法」単複同型
3　　　alternative to 〜：「〜に代わる」
6　　　cutting-edge：⇒ p.9, 8 参照
9　　　that：＝ the environmental problem
11　　generation：⇒ p.11, 22 参照
15　　sustainable：「（資源の利用が）環境破壊をせずに継続できる，（資源が）枯渇することなく利用できる」
17　　to begin with：「まず第一に」
18　　trap：「取り込む」

p.70
9　　　there：＝ in the atmosphere
18　　as yet：「今までのところは」
21　　species：⇒ p.22, 18 参照
24　　starch：「デンプン」
26　　physicochemically：「物理化学的に」
27　　a number of：⇒ p.10, 6 参照

p.71
2　　　amylolytic：「デンプン分解の」
17　　in itself：「それ自体」
18　　bring about：⇒ p.30, 12 参照

Organisms that Eat the Earth

What did you eat today? Was it a salad? A steak? A bowl of rice? What you are actually eating is the sun. But nobody ever gets full just by sunbathing. Only living organisms capable of photosynthesis[*1], namely plants and phototropic bacteria, utilize sunlight for their own growth. Animals and fungi then survive by eating or absorbing the plants that grew through photosynthesis. Even those who say they eat nothing but meat are still eating plants indirectly since they are eating the meat of livestock that grew by eating plants. This connection between the "eaters" and the "eaten" in ecosystems[*2] is referred to as the "food chain" since these relationships extend like the links of a chain. Almost every food chain on earth depends on the photosynthesis of plants : the primary production[*3] from sunlight.

On what, then, do ecosystems depend in the deep-sea[*4], where sunlight fails to reach? Marine organisms also depend on the food chain based on photosynthesis. Since the seas are filled to its depths with saltwater, however, the photic zone[*5]—namely the depth exposed to sunlight sufficient for phytoplankton[*6] to carry out photosynthesis and grow—is 200 meters at most. Although the reproduction of living organisms in the ocean takes place toward the surface depths, not all organisms are consumed through the food chain of the organisms living here : a small amount of uneaten matter, excrement and carcasses sink into the deep sea. These become the food source for the deep-sea ecosystem. To put it another way, there are few organisms able to live at such depths owing to the dearth of food sources in the deep sea far removed from the surface where the reproduction of organisms through photosynthesis takes place.

At the beginning of the 1970s, one of the greatest discoveries of the 20th century in the field of natural history was made on the sea floor at a depth of 2,500 m, off the shores of the Galapagos Islands in the eastern Pacific Ocean. Based on the theory of plate tectonics[7], two plates converge in this area of the ocean and there is a mid-ocean ridge, referred to as the Galapogos Rift, on the ocean floor ; and the eruptions of massive deep-sea volcanoes accompanied by hydrothermal vent from the sea floor were discovered there. Researchers who went underwater in a submersible vessel in order to observe the situation of the sea floor witnessed a surprising spectacle. The sea floor bristled with smokestack-like pipes, which are called "chimneys", venting hot water, and the ocean floor around these were blanketed by living organisms : tube worms[8] that extend a red tongue like appendages into the water from the tip of their thin bodies of over a meter in length ; deep-sea clams[9] which are a phylum of bivalves (*Calyptogena magnifica*) and grow to a size exceeding 10 cm ; and species of small crabs.

In general, about $5 g/m^2$ of biomass[10] is distributed on the sea floor at the depth of 2,500 m. This suggests conditions in which the desert-like ocean floor is sparsely inhabited by living organisms. Yet the biomass of living organisms around the hydrothermal vents comprised several to several tens of kilograms. How were so many living things able to inhabit the deep sea where food is scarce? Here is the reason for this great discovery : Hot water contains sulfur and hydrogen sulfide which chemosynthetic bacteria[11], such as sulfur-oxidizing bacteria and sulfate-reducing bacteria, feed on to reproduce explosively. Crabs and shrimp directly capture and eat these increased colonies of bacteria, and larger benthos[*12], such as tube worms and deep-sea clams, live with chemosynthetic bacteria residing within their bodies (symbiosis[*13]), nourished by the products originating from the bacteria. In this way, the bacteria utilized chemicals dissolved in water to become the primary producer in the food chain, on which all other organisms in this ecosystem rely. In other words, the organisms living in this area are "eating the earth."

Human beings had come to believe that the life of almost all organisms on the face of the earth was sustained by photosynthetic organisms that harness the sun's energy, but this discovery triggered a "paradigm shift," causing us to consider the existence of other types of ecosystems that are sustained by chemosynthetic organisms which utilize chemical substances coming from the inside of the earth.

Other communities of chemosynthetic organisms have been subsequently discovered around converging plates or the ocean floor of geologically active areas. As research has progressed, scientists have come to think that the high-temperature, high-pressure conditions found around hydrothermal vents resemble the primordial environment in which life originated, and those conditions are therefore receiving serious attention to shed some light on the origin of life[*14].

In 2004, a massive methane plume[*15] was discovered on the floor of the Sea of Japan at the depth of 900 meters, off the coast of Joetsu City, Niigata. In the following year, methane hydrate[*16] was discovered on the sea floor through observations using a remotely operated underwater vehicle (ROV)[*17]. Around the methane hydrate were numerous mat-like colonies of methane-oxidizing bacteria obtaining nourishment on methane to reproduce. Moreover, the population of red snow crabs (*Chinoectes japonicus*), highly valued seafood, inhabiting this area was 5 to 120 times higher than usually found on the sea floor of the Sea of Japan. Research is currently being carried out on the relationships among methane, bacteria, and crabs, and should the crabs be found to feed on methane-oxidizing bacteria, all of you who eat snow crabs may also turn out to be a member of the "chemosynthetic ecosystem,"*18 an indirect "eater of the earth."

科学解説

*1　photosynthesis［光合成］

　　　生物が光のエネルギーを利用して二酸化炭素と水から有機物を合成する過程を光合成といいます。光合成の全体の反応は，以下の化学式で表されます。

　　　　$12H_2O(水) + 6CO_2(二酸化炭素) + 光エネルギー \rightarrow C_6H_{12}O_6(グルコース) + 6H_2O + 6O_2(酵素)$

　　　生物が二酸化炭素を取り込み，エネルギーを使って炭水化物などの有機物に作りかえる働きを，炭酸同化といいます。光合成の場合は，炭酸同化に光エネルギーを利用することによって，光エネルギーは有機物中の化学エネルギーに変えられます。したがって，エネルギー転換からみると，光合成は生物界にエネルギーを取り入れる働きをしています。

*2　ecosystem［生態系］

　　　生命圏では，生物は他の生物を食べたり，他の生物から食べられたり，温度や湿度などに影響されたり，常に外から影響を受けながら生活しています。このような生物の生活に何らかの影響を与える外界の要因を「環境」と呼びます。また環境を構成している要因は，光，温度，大気，水などの非生物的（無機的）環境要因と，同種の生物間の関係や異種の生物間の関係，食物資源などの生物的（有機的）環境要因に大別できます。この2つの環境を併せてひとつのまとまったシステムとしてとらえたものを「生態系（エコシステム）」といいます。生態系の範囲は，目的によって任意に決められています。例えば，海洋生態系や森林生態系，極域生態系などといった使われ方をしますが，その境界は曖昧です。

*3　primary production［一次生産］

　　　独立栄養生物による有機物の生産を一次生産といいます。一般的に光合成による有機物の生産がその大部分を占めていますが，硫化水素の存在するような還元環境などでは，化学合成生物が一次生産をする特殊な生態系も存在します。

*4　deep-sea［深海］

　　　太陽光は水中に入ると減衰して（弱くなって），水深が増すにつれて光量が少なくなります。「深海」の定義はいろいろありますが，一般的に植物プランクトンが光合成して成長するのに必要な光が届く水深0mから200mまでの深度帯を有光層といい，有光層以下を「深海」と定義しています。深海の体積は全海洋の約93%にもなります。

*5　photic zone［有光層］

　　　⇒「深海」の解説参照。

*6　phytoplankton［植物プランクトン］

　　　海洋・湖沼・河川などの水中を漂いながら生活する単細胞または群体性の藻類や微小な高等植物。
　　　（⇒ p.66，科学解説*6参照）

*7　plate tectonics［プレートテクトニクス（理論）］

　　　地球の表面は厚さ約100kmの固い層（リソスフェア）からできていて，それはさらにプレートと呼ばれるいくつかの大規模な岩盤に分かれています。プレートの境界は，地球内部のマントルの一部が湧き上がってきてプレートが作り出されている「発散境界」と，プレートが地球内部へもぐり込む「収集境界」，プレートどうしがすれ違う「横ずれ境界」に分けられます。プレートは発散境界で作り出され，

年間1〜10cm程度の速さで水平方向へ移動し，収束境界で他のプレートと衝突して地球内部へもぐり込んで行きます。このプレートの移動が，大陸移動など地球規模の地殻変動を起こす原動力となっているという理論です。

*8　tube worm［チューブワーム］

　　ゴカイ（多毛類）の仲間で，細長い管の中に生物体が入っています。日本では「ハオリムシ」の仲間とされています。長さは種によって異なり，数十cmから2m。白い外筒の先端から鰓（えら）突起を出して，管の中の体は軟らかく，大半をソーセージ状の栄養体が占めています。口も消化管もなく，栄養体に化学合成細菌を共生させていて，栄養を供給させています。

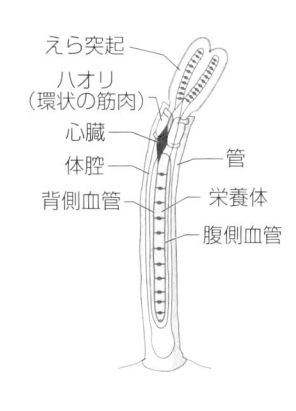

ハオリムシのしくみ

*9　deep-sea clam［シロウリガイ類］

　　オトヒメハマグリ科に属す二枚貝の仲間で，殻の大きさは種によって異なりますが，殻長1〜30cmになります。シロウリガイ類は鰓（えら）が発達していて，鰓に硫黄酸化細菌が共生しています。消化器官は著しく退化していて，チューブワームと同様に，化学合成細菌に栄養を供給させています。

*10　biomass［バイオマス］

　　生物体量，生物量，現存量とも呼ばれます。生態学の分野で「バイオマス」とは，ある時点に任意の空間内に存在する生物体の量を，重量ないしエネルギー量で示した指標のことをいいます。

*11　chemosynthetic bacteria［化学合成バクテリア］

　　無機物や単純な有機物の酸化によって発生するエネルギーを利用して，炭酸固定して有機物を合成するバクテリア（細菌）です。（⇒ p.80．科学解説 *18 参照）

*12　benthos［底生生物］

　　海洋・湖沼・河川などの水域にすむ生物のうち，貝類や多毛類（ゴカイ類），甲殻類（エビ類・カニ類など），海藻類（コンブやワカメなど）など，水底に生活する生物の総称です。

*13　symbiosis［共生］

　　異種生物の個体または個体群（同種個体の集まり）が，行動的または生理的に密接な関係を保ちながら一緒に生活している現象を共生といいます。共生は，相利共生（双方が明らかに利益を交換している共生関係）・片利共生（片方だけが利益を得ているが，もう片方には影響しない共生関係）・寄生（片方がその栄養を他方の生物体の一部からとって生活するなど，片方の生物に害を及ぼす関係）の3つに大別できます。

*14　origin of life［生命起源］

　　生命は，原始の地球上に存在した簡単な化学物質や隕石に含まれていたアミノ酸などが化学反応を繰り返すことにより，約38億年前に発生したとされています。化学反応を促進させるためには，熱や放電などのエネルギーが必要です。深海では，水圧のため水は100℃ではなく，300℃以上で沸騰することがあります。深海の海底火山の火口周辺は，非常に高温になり化学反応が促進される可能性があることなどから，生命の起源を研究するうえで注目されています。

＊15　plume［プルーム］

　　　プルームとは「羽毛」のことです。海底からメタンなどが吹き出していると，ちょうど煙突から煙が出るように吹き出し口は狭いのですが，吹き出した物質は海中に広がりながら浮上していきます。この状態が1本の羽毛を横にした形のように見えるので「プルーム」といいます。

＊16　methane hydrate［メタンハイドレート］

　　　ハイドレートとは「水和物」のことで，水分子はある温度・圧力環境では分子レベルのかご状の構造を作ります。そのかご構造の中にメタン分子が含まれているものをメタンハイドレートと呼びます。メタンハイドレートは氷状になり，燃えるので「燃える氷」とも呼ばれています。メタンハイドレートは，その体積の160～170倍のメタンを含んでいます。

＊17　remotely operated underwater vehicle（ROV）［遠隔操作無人探査機］

　　　潜水調査船のうち，海上などの船舶から遠隔操作によってコントロールするものをいいます（remotely operated vehicle）。これに対して有人潜水調査船はHOV（human operated vehicle）といいます。

＊18　chemosynthetic ecosystem［化学合成生態系］

　　　化学合成とは無機物や単純な有機物の酸化によって発生するエネルギーを利用して炭酸を固定して有機物を合成することをいいます。無機物としては，硫化水素・硫黄・アンモニア・亜硝酸・水素・鉄などが利用されて，それらの酸化エネルギーから有機物を合成できる生物は化学合成無機栄養生物といい，硫黄細菌・硝化細菌・水素細菌・鉄細菌などが含まれます。また，メタン酸化細菌は，単純な有機物であるメタンをエネルギーや炭素源として生育するので，化学合成有機栄養生物といいます。これら2つのタイプの化学合成生物は，化学エネルギーを利用して，生体物質を作り上げることから，生態系内の生産者的な役割をしています。そして化学合成生物が光合成生物の代わりに生産者的に機能する生態系を「化学合成生態系」と呼びます。しかし，化学合成生態系は，光合成生態系から独立しているわけではなく，化学合成の過程に必要な酸素や動物が呼吸するための酸素は，光合成の過程で生産された酸素を利用しています。

英語 註

p.74

行	
2	full：p.35, 2参照
4	phototropic bacteria：「光合成細菌」
5	fungus：「菌類」複数形はfungi
17	at most：⇒ p.63, 27参照
17～18	take place：⇒ p.11, 10参照
20	excrement：「糞」
20	carcass：「死骸」
21	to put it another way：「言い換えれば」

p.75

3	take place：⇒ p.11, 10 参照
6	the Galapagos Islands：「ガラパゴス諸島」（太平洋の赤道直下にあるエクアドル領の火山島群，珍しい動物に富み，チャールズ・ダーウィンが進化論のヒントを得た島として有名）
10	hydrothermal vent：「熱水噴出孔」
13	bristle with ～：「～でいっぱいである，～が密生する」
15	blanket：「一面に覆う」
23～24	several to several tens of kilograms：「数キロから数十キロ」
26	hydrogen sulfide：「硫化水素」
27	sulfur-oxidizing bacterium：「硫黄酸化バクテリア」
27	sulfate-reducing bacterium：「硫酸還元バクテリア」

p.76

11	paradigm shift：「パラダイム・シフト」（ものの見方を根本的に規定している概念的枠組みが変わること）
27	methane-oxidizing bacterium：「メタン酸化バクテリア」

p.77

5	should the crabs be found：if the crabs should be found「万一～ならば」
6	turn out to be ～：「～であることがわかる」

科学解説　索引

acetyl CoA：アセチル CoA ………………………………………………………………… *44*
adjuvant effect：免疫賦活効果 …………………………………………………………… *32*
aerosol：エアロゾル ……………………………………………………………………… *66*
allergen：アレルゲン ……………………………………………………………………… *32*
AMPK：AMP 依存性キナーゼ …………………………………………………………… *38*
Antarctic krill：ナンキョクオキアミ …………………………………………………… *67*
antibody：抗体 …………………………………………………………………………… *32*
appetite center：食欲中枢 ………………………………………………………………… *38*
arteriosclerosis：動脈硬化症 ……………………………………………………………… *50*
ATP（adenosine triphosphate）：アデノシン三リン酸 ………………………………… *38*
bacteriocin：バクテリオシン ……………………………………………………………… *13*
bacteriophage：バクテリオファージ …………………………………………………… *13*
benthos：底生生物 ………………………………………………………………………… *79*
bioethanol：バイオエタノール …………………………………………………………… *72*
biomass：バイオマス ……………………………………………………………………… *79*
blood glucose level：血糖値 ……………………………………………………………… *38*
carbon negative：カーボンネガティブ ………………………………………………… *73*
carbon neutral：カーボンニュートラル ………………………………………………… *72*
cell cycle：細胞周期 ……………………………………………………………………… *60*
chemosynthetic bacteria：化学合成バクテリア ………………………………………… *79*
chemosynthetic ecosystem：化学合成生態系 …………………………………………… *80*
deep-sea：深海 …………………………………………………………………………… *78*
deep-sea clam：シロウリガイ類 ………………………………………………………… *79*
diabetes：糖尿病 …………………………………………………………………………… *55*
dimethyl sulfide（DMS）：硫化ジメチル ……………………………………………… *66*
epigenetic：エピジェネティク …………………………………………………………… *60*
esophagus：食道 …………………………………………………………………………… *25*
familial tumor：家族性腫瘍 ……………………………………………………………… *60*
food chain：食物連鎖 …………………………………………………………………… *67*
fossil fuel：化石燃料 …………………………………………………………………… *72*
free fatty acid：遊離脂肪酸 ……………………………………………………………… *38*
gene recombination technology：遺伝子組換え技術 …………………………………… *13*
global warming：地球温暖化 …………………………………………………………… *72*
gluconeogenesis：糖新生機構 ……………………………………………………… *38, 44*
glucose：グルコース ……………………………………………………………………… *44*
glycogen：グリコーゲン …………………………………………………………………… *44*
greenhouse effect：温室効果 ……………………………………………………………… *66*
greenhouse gas：温室効果ガス …………………………………………………………… *66*
helper T lymphocyte：ヘルパー T 細胞 ………………………………………………… *32*
heterozygous defect：ヘテロ接合性欠損 ………………………………………………… *60*
hormone：ホルモン ……………………………………………………………………… *38*
hydrolyze：加水分解する ………………………………………………………………… *44*
ice sheet：氷床 …………………………………………………………………………… *66*
inflammatory bowel disease：炎症性腸疾患 …………………………………………… *32*
insulin：インスリン ……………………………………………………………………… *55*
insulin receptor：インスリン受容体 …………………………………………………… *55*

intestinal microbiota：腸内細菌叢	32
IPCC（Intergovernmental Panel on Climate Change）：気候変動に関する政府間パネル	66
islets of Langerhans：ランゲルハンス島	55
ketoacidosis：ケトアシドーシス	55
ketone body：ケトン体	55
key species：鍵種	67
lactic acid bacteria：乳酸菌	13
large intestine：大腸	25
metabolic syndrome：メタボリックシンドローム	50
metabolome analysis：メタボローム解析	44
methane hydrate：メタンハイドレート	80
mouth：口腔	25
mutant：変異体	60
nervous histamine：神経ヒスタミン	39
neutral lipid：中性脂肪	44
nutrient：栄養素	26
obese gene（*ob* gene）：肥満遺伝子	38
oleuropein：オレウロペイン	19
oral vaccine：経口ワクチン	33
origin of life：生命起源	79
oxidize：酸化する	44
pack ice area：流氷域	67
peptidase：ペプチダーゼ	13
photic zone：有光層	78
photosynthesis：光合成	78
phytoplankton：植物プランクトン	66, 78
plate tectonics：プレートテクトニクス理論	78
plume：プルーム	80
polyphenol：ポリフェノール	19
primary production：一次生産	78
promoter：プロモーター	13
protein：タンパク質	26
reactive oxygen species（ROS）：活性酵素	19
reduce：還元する	44
remotely operated underwater vehicle（ROV）：遠隔操作無人探査機	80
salpa：サルパ	67
small intestine：小腸	25
stomach：胃	25
substrate specificity：基質特異性	13
symbiosis：共生	79
thrifty gene：倹約遺伝子	50
tube worm：チューブワーム	79
α-lipoic acid：α-リポ酸	39

編修	寺本 明子	東京農業大学 応用生物科学部 准教授
	James W. Hove	
	田所 忠弘	東京農業大学 応用生物科学部 教授
共著	小林 謙一	東京農業大学 応用生物科学部 助教
	佐藤 英一	東京農業大学 応用生物科学部 准教授
	沼波 秀樹	東京家政学院大学 現代生活学部 准教授
	山本 祐司	東京農業大学 応用生物科学部 教授

Learning Life Science in English
英語で読み解く生命科学

2012年（平成24年）3月20日　初版発行

編　修　寺本 明子
　　　　James W. Hove
　　　　田所 忠弘
発行者　筑紫 恒男
発行所　株式会社 建帛社
　　　　KENPAKUSHA

〒112-0011　東京都文京区千石4丁目2番15号
　　　　　　TEL （03）3944-2611
　　　　　　FAX （03）3946-4377
　　　　　　http://www.kenpakusha.co.jp/

ISBN978-4-7679-4638-2　C3082　　　　　　教文堂／プロケード
©寺本明子ほか，2012.　　　　　　　　　　　　Printed in Japan
（定価はカバーに表示してあります）

本書の複製権・翻訳権・上映権・公衆送信権等は株式会社建帛社が保有します。
JCOPY　〈(社)出版者著作権管理機構　委託出版物〉
本書の無断複写は著作権法上での例外を除き禁じられています。複写される場合は，そのつど事前に，(社)出版者著作権管理機構（TEL 03-3513-6969，FAX 03-3513-6979，e-mail:info@jcopy.or.jp）の許諾を得て下さい。